Advances in Anatomy, Embryology and Cell Biology publishes critical reviews and state-of the-art surveys on all aspects of anatomy and of developmental, cellular and molecular biology, with a special emphasis on biomedical and translational topics.

The series publishes volumes in two different formats:

- Contributed volumes, each collecting 5 to 15 focused reviews written by leading experts
- Single-authored or multi-authored monographs, providing a comprehensive overview of their topic of research

227
Advances in Anatomy, Embryology and Cell Biology

More information about this series at
http://www.springer.com/series/102

Julita Kulbacka • Saulius Satkauskas
Editors

Transport Across Natural and Modified Biological Membranes and its Implications in Physiology and Therapy

Springer

Editors
Julita Kulbacka
Wroclaw Medical University
Dept. of Medical Biochemistry
Wroclaw
Poland

Saulius Satkauskas
Vytautas Magnus University
Bio. Res. Group/Bio Dept.
Kaunas
Lithuania

ISSN 0301-5556 ISSN 2192-7065 (electronic)
Advances in Anatomy, Embryology and Cell Biology
ISBN 978-3-319-56894-2 ISBN 978-3-319-56895-9 (eBook)
DOI 10.1007/978-3-319-56895-9

Library of Congress Control Number: 2017954301

Printed on acid-free paper

This Springer imprint is published by Springer Nature
The registered company is Springer International Publishing AG
The registered company address is: Gewerbestrasse 11, 6330 Cham, Switzerland

Preface

The physiology of the living cells strongly depends on the cell membranes and the molecules' transportation. Biological membranes are dynamic layered structures composed of phospholipids and proteins. These structures delimit separate compartments in biological systems to maintain separation and integrity of the cell. Membranes maintain concentration gradient of various substances and the electrical potential difference between the inside and outside of the cell. The membranes are in fact specialized in transport systems, allowing for the controlled flow of various molecules and ions. Transportation is necessary for cells in order sustain all life processes. Biological membranes are responsible for cell communication and recognition and participate in the metabolic reaction of the cell. This natural barrier toward various molecules is also a problem in the transport and drug delivery to cells.

The current volume entitled *Transport Across Natural and Modified Biological Membranes and Its Implications in Physiology and Therapy* presents insights stimulating the continuing efforts to understand mechanisms involved in the biological membrane field. This volume contains the knowledge of the new modalities and characterization for basic in vitro and computer models of biological membranes; analysis in terms of advances in molecular dynamics, in-depth analysis of images from various biological models; analysis of membrane models for treatment trials; and new insights and current concepts in overcoming biological membranes in treatment procedures (electroporation, sonoporation, channel blockers, etc.).

This book will guide the reader from the recent advances of atomistic modeling of model cell membranes and electroporation. Computer modeling assists in elucidating the structure, membrane functions, and transport mechanisms. Continuing with a focus on computer modeling, authors presented the role of bioinformatics in the ion channel study. Transport mechanisms existing in the biological membranes highly determine proper cellular functions and contribute to drug transport. Authors summarized the current knowledge on features and electrical properties of the cell membrane and described how the cell membrane accomplishes transport functions and how transmembrane transport can be affected. The changes occurring in the cell membrane during electroporation or electropermeabilization in the context of chemical analysis of cell membrane modifications and associated transport mechanisms are also discussed. Physical methods for drug and gene delivery through the cell plasma membrane are presented.

Cell membranes contain various specialized protein molecules, i.e., membrane receptors. Their function is crucial because external molecules attach, triggering changes in the cell function. Thus, in one of the chapters, the regulation and signaling of estrogen receptors in cell membranes are reviewed. The next chapter deals with the issue involved in imaging of membrane and electroporation and on how processes can improve its development in cell biology and clinics. Most of our knowledge about microscopic structures including cells and membranes is conveyed via images. In the last chapter, selected methods of analysis of images of cells and biological membranes such as detection, segmentation, classification and machine learning, registration, tracking, and visualization are reviewed. Authors indicate that the detailed analysis of membrane images could facilitate understanding of the underlying physiological structures or help in the interpretation of biological experiments.

The scientific knowledge reviewed in this volume promotes the research involved on natural and modified biological membranes, transport across the membrane barrier, and methodology associated with membrane feature modulation and finally their analysis.

Wroclaw, Poland Julita Kulbacka
Kaunas, Lithuania Saulius Satkauskas

Acknowledgment

I would like to thank all authors – great scientists who decided to contribute to this book project. I would like also to express my appreciation to their knowledge, effort, and priceless discussions that allowed us to publish this book. Moreover, we'd like to give special thanks to Professor Horst-Werner Korf for his support and valuable comments. I hope our chapters will spread and clarify the knowledge about the transport across natural and modified biological membranes and its implications in physiology and therapy.

Contents

Chapter 1
Atomistic Simulations of Electroporation of Model Cell Membranes

Mounir Tarek

Abstract Electroporation is a phenomenon that modifies the fundamental function of the cell since it perturbs transiently or permanently the integrity of its membrane. Today, this technique is applied in fields ranging from biology and biotechnology to medicine, e.g., for drug and gene delivery into cells, tumor therapy, etc., in which it made it to preclinical and clinical treatments. Experimentally, due to the complexity and heterogeneity of cell membranes, it is difficult to provide a description of the electroporation phenomenon in terms of atomically resolved structural and dynamical processes, a prerequisite to optimize its use. Atomistic modeling in general and molecular dynamics (MD) simulations in particular have proven to be an effective approach for providing such a level of detail. This chapter provides the reader with a comprehensive account of recent advances in using such a technique to complement conventional experimental approaches in characterizing several aspects of cell membranes electroporation.

1.1 Introduction

Research in biochemistry and molecular biology of bio-membranes has experienced remarkable progress in the last 30 years, particularly with the realization that many different classes of lipids play fundamental roles in the structure and organization of the cell membranes. Accordingly, the search for new cell membrane manipulation techniques to design novel therapies has triggered interest in combining engineering technologies with medical and biological knowledge and became the basis for the development of innovative and effective treatments of various diseases.

M. Tarek
CNRS, Université de Lorraine, F-54506 Vandoeuvre les Nancy, France
e-mail: mounir.tarek@univ-lorraine.fr

J. Kulbacka, S. Satkauskas (eds.), *Transport Across Natural and Modified Biological Membranes and its Implications in Physiology and Therapy*, Advances in Anatomy, Embryology and Cell Biology 227,
DOI 10.1007/978-3-319-56895-9_1

Electroporation (Neumann et al. 1989) stands out as the probably oldest of such techniques. Based on the effects of external pulsed electric fields, it enhances the permeability of cell culture media and tissues to otherwise poorly or non-permeant species (Benz et al. 1979; Weaver and Chizmadzhev 1996; Chen et al. 2006). In fact, electroporation transiently or permanently disturbs the integrity of cell membranes as manifested by a substantial increase of the transmembrane ionic conductance or by an increase of molecular uptake or release from the cell (Mir et al. 1988; Hibino et al. 1991; Teissié et al. 1999; Pucihar et al. 2008; Breton et al. 2012). Today, the technique is widely used in biomedicine and biotechnology (Yarmush et al. 2014; Teissié et al. 1999; Villemejane and Mir 2009; Breton and Mir 2012; Teissie 2013; Miklavčič et al. 2014; Cadossi et al. 2014; Lakshmanan et al. 2014). Its most impressive biomedical applications range from in vitro DNA and siRNA cell delivery to clinical electro-chemotherapy.

In standard electroporation (EP), the electric pulses are of μs ÷ ms time scale and their amplitude in the order of kV/cm (pulses electric fields called hereafter msPEFs). The effect of msPEFs is to increase the local transmembrane voltage that, when exceeding a certain threshold, to produce an intense local electric field at the membrane level leading to pore formation (Kotnik et al. 2010). In recent years, pulse generators that produce ultrashort (ns) intense (tens of kV/cm) electric fields (nsPEFs) have been designed and shown not only to induce similar effects to classical EP but also opened new perspectives of cell electro-manipulation because of their capability to also permeabilize the membranes of internal organelles (Deng et al. 2003; Vernier et al. 2006a; Chopinet and Rols 2015).

Depending on the cell type, function, and species, the thickness of biological membranes may vary from approximately 2–10 nm. Regardless of their tasks, all biological membranes have, however, a number of common characteristics: they all consist of an assembly of lipids, proteins, and carbohydrates that self-organize into a thin barrier that separates the interior of cell compartments from the outside environment (Gennis 1989). Furthermore, the main lipid constituents of all cell membranes are phospholipids, molecules of polar or hydrophilic head and nonpolar or hydrophobic tails formed of two fatty acids that arrange themselves into a two-layered sheet (a bilayer). Experimental evidence suggests that under electroporation conditions, the applied electrical pulse induces a large change in the transmembrane voltage and rearrangements of the membrane components, in particular of the lipid bilayer leading to the formation of aqueous pores (Abidor et al. 1979; Benz et al. 1979; Weaver and Chizmadzhev 1996; Weaver 2003; Chen et al. 2006; Tarek 2005; Szabo and Wallace 2015; Kotnik et al. 2010).

Because of the remarkable fluidity of membranes under physiological conditions, the knowledge of both their structural and dynamical changes induced by electric fields remains to a large extent fragmentary. The key features of electroporation are based on theories involving stochastic pore formation. Although extensive experimental work on cells and model lipid membranes (e.g., planar lipid bilayers and lipid vesicles) has been devoted to the structural characterization of pores (Kalinowski et al. 1998; Koronkiewicz et al. 2002; Koronkiewicz and Kalinowski 2004; Pavlin et al. 2005; Kotulska 2007; Krassen et al. 2007; Kramar et al. 2012; Bowman et al. 2010;

Nesin et al. 2011; Silve et al. 2012; Pakhomov et al. 2015), the data emerging so far is very scarce. Only very recently, time-resolved visualization of pores in bilayers was made possible using total internal reflection fluorescence microscopy (Szabo and Wallace 2015). Nevertheless, such imaging currently does not allow the spatial resolution required to measure the pore sizes, neither their short time scale dynamics.

1.2 Molecular Dynamics Simulations of Lipid Membranes

MD simulations are based on molecular mechanics and a classical treatment of particle-particle interactions that precludes bond dissociation and therefore the simulation of chemical reactions (Allen and Tildesley 1987; Leach 2001). They require the choice of a potential energy function, i.e., terms by which the particles interact, usually referred to as a force field. Those most commonly used in biophysics, e.g., GROMOS (Schuler et al. 2001), CHARMM (Klauda et al. 2010), and AMBER (Case et al. 2008) consist of a summation of bonded forces associated with chemical bonds, bond angles, and bond dihedrals, and non-bonded forces associated with van der Waals forces and electrostatic interactions. The parameters describing these terms are optimized to reproduce structural and conformational changes of macromolecular systems.

MD simulations use information (positions, velocities or momenta, and forces) at a given instant in time, t, to predict the positions and moment at a later time, $t + \Delta t$, where Δt is the time step, of the order of a femtosecond, taken to be constant throughout the simulation. Numerical solutions to the equations of motion are thus obtained by iteration of this elementary step using computers. Being limited by the speed of execution of the programs, and the availability of computer power, simulations are usually performed on a small number of molecules (few tens to few hundred thousand atoms). In order to eliminate edge effects and to mimic a macroscopic system, simulations of condensed phase systems consider such small patches as confined in a central simulation cell and replicate the latter using periodic boundary conditions (PBCs) in the three directions of the Cartesian space.

Phospholipids have served as models for investigating in silico the structural and dynamical properties of lipid membranes. Though zwitterionic phosphatidylcholine (PC) lipid bilayers constituted the best characterized systems during the last two decades (Chiu et al. 1999; Feller et al. 2002; Rög et al. 2002; Saiz and Klein 2002), more recent studies have considered a variety of alternative lipids (Berkowitz and Raghavan 1991; Damodaran and Merz 1994; Cascales et al. 1996; Chiu et al. 2003; Mukhopadhyay et al. 2004), or mixed compositions (Pandit et al. 2003; Dahlberg and Maliniak 2008; Gurtovenko and Vattulainen 2008; Patel and Balaji 2008; Li et al. 2009; Rog et al. 2009; Vacha et al. 2009). Despite their simplicity, bilayers built from such lipids represent remarkable test systems to probe the computation methodology and to gain insight into the physical properties of membranes (Tobias et al. 1997; Tarek et al. 2001; Anézo et al. 2003; Chipot et al. 2005). Classical force fields, which are undergoing constant improvement, were shown to provide an adequate description of the properties of membrane systems.

1.3 Modeling Lipid Membrane Electroporation

When cells are subject to msPEFs by placing them, for instance, between two electrodes and applying a constant voltage pulse, the resulting current causes accumulation of ionic charges at both sides of the cell membrane. The charging time constant is dependent upon the electrical parameters of the medium in which the cell is suspended and is in the range of hundreds of nanoseconds (Pauly and Schwan 1959; Kotnik et al. 1998; Kotnik and Miklavcic 2006). If, on the other hand, the pulse duration is short enough relative to the charging time constant of the resistive-capacitive network formed by the conductive intracellular and extracellular fluids and the cell membrane dielectric, which is the case for nsPEFs, then the response of the system is mainly dielectric and is linked to the polarization of the interfacial water. Simulation protocols have therefore been devised to perform in silico experiments under both conditions, i.e., submitting lipid bilayers either to megavolt per meter pulsed electric fields or to a charge imbalance, mimicking therefore the application of low voltage – long duration pulses (Casciola and Tarek 2016).

In simulations of lipid bilayers, an applied electric field $\vec{E}$ (in practice, this is done by adding a force $\vec{F} = q_i\vec{E}$ to all the atoms bearing a charge q_i) acts mostly on the interfacial water dipoles (very small polarization of bulk water). The reorientation of the lipid head-groups is hardly affected at very short time scales (Tarek 2005; Vernier and Ziegler 2007) and does not exceed few degrees toward the field direction at longer time scale (Böckmann et al. 2008). Within typically few picoseconds (Tarek 2005; Delemotte and Tarek 2012), a transverse field $\vec{E}$ induces an overall transembrane potential $\Delta V \approx \left|\vec{E}\right|.L_Z$ where L_Z is the size of the simulation box in the field direction due to the MD setup and the use of PBCs. In the example shown in Fig. 1.1, L_Z is ~ 10 nm. Accordingly the electric field applied (0.2 V.nm^{-1}) induces ΔV ~ 2V across the POPC bilayer.

The application of an electric field of such high enough magnitude in MD simulations of pure lipid bilayers leads to membrane electroporation (Tieleman 2004; Hu et al. 2005; Tarek 2005; Böckmann et al. 2008; Ziegler and Vernier 2008). This process involves a common poration sequence: The electric field favors quite rapidly (within a few hundred picoseconds) formation of water defects and water wires deep into the hydrophobic core (Tieleman 2004). Ultimately water fingers forming at both sides of the membrane join up to form water channels (often termed prepores or hydrophobic pores) that span the membrane. Within nanoseconds, lipid head-groups start to migrate from the membrane-water interface to the interior of the bilayer, stabilizing a hydrophilic pore of ~1 to 5 nm diameter. All MD studies report pore expansion when the electric field is maintained. Systematic studies of pore creation and annihilation life time as a function of field strength have shed more light onto the complex dynamics of pores in simple lipid bilayers (Böckmann et al. 2008; Levine and Vernier 2010, 2012). Hence, while pioneering simulations allowed for the first time to confirm the electroporation theories involving stochastic pore formation and to "visualize" the event, subsequent state-of-the-art MD simulations of lipid bilayers subject to electric fields have provided a wealth of molecular level details of the phenomena.

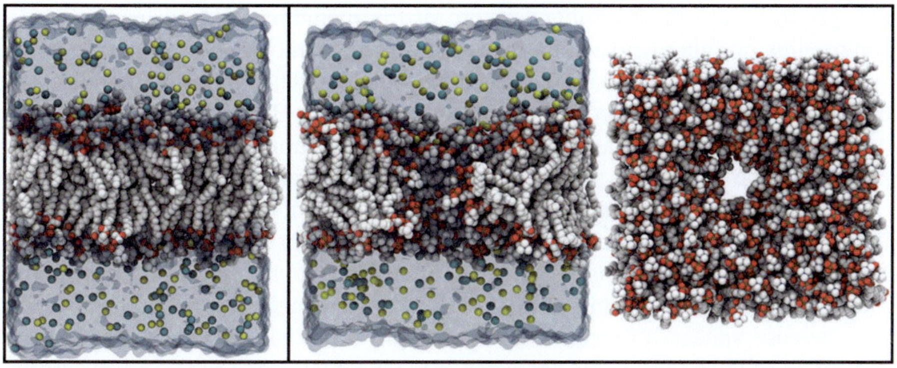

Fig. 1.1 *Left*: Configuration of a palmitoyl-oleoyl-phosphatidylcholine (POPC) hydrated bilayer system from a well-equilibrated constant pressure MD simulation performed at 300 K. *Right*: Side and top views of the electroporation of the bilayer from MD simulations carried out at ~500 mV, using the modified protocol (Casciola et al. 2016). Only the molecules in the simulation cell are shown. The phosphate (*blue*) and nitrogen (*purple*) atoms of the lipid head-groups, the carbonyl oxygen (*red*), and the carbon atoms of the lipid acyl chains (*white*) are depicted by their van der Waals radius. Ions (sodium in *yellow* and chloride in *cyan*) of the extra- and intracellular medium are also represented by their van der Waals radius, while water molecules are not shown for clarity and replaced by a transparent cyan shadow. Note the formation of a large water pore stabilized by the lipid head-groups

Due to the use of 3d PBCs, the transmembrane (TM) voltage cannot be controlled by imposing, in a classical setup of membrane simulations, a net charge imbalance (Q_s) across the bilayer. TM potential gradients can, however, be generated by Q_s across lipid bilayers by considering an MD unit cell consisting of three saltwater baths separated by two bilayers and 3d-PBCs (Sachs et al. 2004). A variant of this method consists in considering a unique bilayer surrounded by solution baths, each of them terminated by an air/water interface (Pandit et al. 2003) maintaining therefore a separation between the upper and lower electrolytes. Q_s is in this case generated by simply displacing at time $t = 0$ an adequate number of ions from one side to the other.

Using the charge imbalance setup, it was possible to directly demonstrate in silico that lipid bilayers behave as capacitors (Delemotte et al. 2008; Delemotte and Tarek 2012). Indeed, simulations of the same lipid bilayer subject to various Q_s showed a linear variation of the transmembrane voltage ΔV induced by such charge imbalances, from which the capacitance could be estimated as $C = Q_s . \Delta V^{-1}$. The capacitance values extracted from simulations depend on the lipid composition (charged or not) and on the force field parameters used and as such, constitute a way of checking the accuracy of lipid force fields used in the simulation. In the case of POPC bilayers embedded in a 1 M solution of NaCl, $C \sim 0.85\ \mu F.cm^{-2}$ which is in reasonable agreement with the value usually assumed in the literature, e.g., 1.0 μF. cm^{-2} (Roux 1997; Sachs et al. 2004) and with measurements for planar POPC lipid bilayers in a 100 mM KCl solution (0.5 μF.cm^{-2}) (Kramar et al. 2012). As in the case of the electric field method, for ΔV above 1.5–2.5 Volts, the electroporation process induced by a net charge imbalance starts with the formation of water fingers that protrude into the hydrophobic core of the membrane. Within nanoseconds, water wires bridging between the two sides of the membrane under voltage stress appear. When the simulations are further expended, lipid head-groups migrate along one wire and form a hydrophilic connected pathway. Because salt solutions are explicitly considered in these simulations, ion conduction through the hydrophilic pores could be monitored, and details about the ionic transport could be gathered.

Hence, simulations have revealed that, regardless of the method used, charge imbalance or direct electric field, i.e., for both msPEFs or nsPEFs protocols used in electroporation-based technologies and treatments, lipid bilayer's permeability increases substantially within identical time scales (namely, few nanoseconds), and this increase appears to result from similar physical changes initiated by a local disruption of the hydrophobic character of the membrane due to water penetration. Indeed, it was shown that water molecules initially restrained to the hydrophilic interfacial region tend to orient their dipoles along the local electric field created by the TM voltage ΔV and form small wires and clusters through intermolecular hydrogen bonds. Such structure can overcome the free energy barrier of penetrating the lipid hydrophobic core as the electric field (voltage) is maintained, "grow" from both sides of the membrane until merging to form water columns (often termed pre-pores or hydrophobic pores) that span the membrane (Tarek 2005; Tarek and Delemotte 2010; Delemotte and Tarek 2012; Casciola and Tarek 2016).

1.3.1 Electroporation Thresholds

A large variety of lipid bilayers has been modeled in order to determine the key elements that modulate their EP thresholds. The increase of such a threshold upon addition of cholesterol (Koronkiewicz and Kalinowski 2004; Kakorin et al. 2005; Casciola et al. 2014) was, for instance, studied using the E field (Fernández et al. 2012) and charge imbalance protocols and was linked to the increase of the stiffness of the bilayer (Koronkiewicz and Kalinowski 2004; Kakorin et al. 2005). Tarek's group investigated the effect on the EP threshold of ester and ether linkages of branched (phytanoyl) tails and of bulky (glucosyl-myo and myo-inositol) lipid head-groups (Polak et al. 2013, 2014). It was then shown that the EP threshold of a lipid bilayer depends not only on the "electrical" properties of the membrane, i.e., its dipole potential or membrane capacitance, but also on the nature of lipids´ hydrophobic tails. The authors reported a correlation between the EP threshold and the lateral pressure exerted at the water/lipid interface, which hinders the local diffusion of water molecules toward the interior of the hydrophobic core, lowering therefore the probability of pore formation.

Comparing specifically the Archea lipids (glucosyl-myo and myo-inositol head-groups) to normal PC lipid, the higher electroporation thresholds for the former was attributed to the strong hydrogen-bonding network stabilizing the lipid head-group interactions (Polak et al. 2013, 2014). A higher EP threshold for phosphatidyletha-nolamine (PE) lipid bilayers compared to phosphatidylcholine (PC) lipid bilayers was also linked to inter-lipid hydrogen bonding taking place in the PE bilayer (Gurtovenko and Lyulina 2014). Studies of more complex composition membranes (Piggot et al. 2011) reported that the cell membrane of Gram-positive bacteria *S. aureus* is less resistant to electroporation than that of the Gram-negative bacteria *E.coli*, a property that was linked to the reduced mobility of the lipopolysaccharide molecules that are located in its outer leaflet.

1.3.2 Pores Features

While most MD simulations support the hypothesis that electroporation leads to the formation of conducting hydrophilic pores stabilized by the lipid head-groups, some studies challenged this fact. Tarek and coauthors pointed out that a peculiar EP process may be possible in which large long-living ion-conducting water columns are not stabilized by lipid head-groups. The migration of lipids along the water column turns out to be largely hindered for palmitoyl-oleoyl-phosphatidylserine (POPS) bilayers characterized by negatively charged head-groups (Dehez et al. 2014). Similar conclusions were drawn for PC lipid bilayers containing more than 30 mole % cholesterol (Casciola et al. 2014) or for Archaea lipids (Polak et al. 2014). Such pore morphologies were ascribed to the repulsion of negatively charged head-groups in the first case, to the condensing effect of cholesterol in the second, and to the steric hindrance of the bulky head-groups coupled with the branched tails in the last case.

1.3.3 Pores Conduction

Using the electric field protocol, it is possible to stabilize the size of the pore to a few nanometers, by lowering the field once the pore is formed (Böckmann et al. 2008; Fernández et al. 2012). Our group (Casciola et al. 2016) used a novel scheme to maintain a constant charge imbalance, which refined the initial μs ÷ ms PEFs approach to model membranes under constant voltages and therefore led to steady pores. In this procedure, the number of ions in the outer and inner electrolytes are frequently estimated and reset by a swapping (Kutzner et al. 2011) to maintain the charge imbalance even when conduction through the pore takes place.

Using the nsPEF- (Fernández et al. 2012; Ho et al. 2013) and μs ÷ msPEF (Casciola et al. 2016)-modified protocols allowed to better characterize the conductance of electropores of various lipid bilayers. For the hydrophilic pores, the radius, and the conductance were found to vary almost linearly with the applied voltage. Moreover, the pores were found to be more selective to cations than to anions (Leontiadou et al. 2007; Ho et al. 2013; Casciola et al. 2016). This selectivity arises from the nature of the lipid molecules constituting the pore. The negatively charged phosphate groups that form the walls of the pore attract cations, which hinders their passage across the bilayer, but also makes the pore interior electrostatically unfavorable for other cations (Gurtovenko and Vattulainen 2007).

The first attempt to link the experimental evidence of pore conductance and the radius estimation was carried out using a linear rising current technique on planar POPC bilayers combined with MD simulations performed under similar conditions (Kramar et al. 2012). It was then suggested that the opening and closing of a single pore under conductance in the 100-nS scale would be possible for a pore diameter of ~5 nm. Extensive simulations of EP of the same bilayer embedded in 1 M NaCl using the modified protocols (Casciola et al. 2016) show that hydrophilic pores with stable radius (1–2.5 nm) form under transmembrane voltages ranging from ~400 to 650 mV, and allow for the conductance in the range of 6.4–29.5 nS (Casciola et al. 2016). Quite interestingly, when the same TM voltages were maintained on the lipid patch of the sizes usually employed in MD simulations (~8.8×8.8 nm^2, 256 lipids), the pore radius obtained were twice smaller, and the conductance almost three times smaller than those reported for the large system (~17.8×17.8 nm^2, 1024 lipids), likely a consequence of the constraints imposed by the finite size of the simulation box.

1.3.4 Transport of Molecules Across Electropores

Although the electroporation-based technologies and treatments concern mainly transport of molecules, e.g., dyes, drugs, and genetic material, across permeabilized cell membranes, the use of MD simulations to investigate such processes is in its infancy. Apart from studies reporting electropore-mediated flip-flop of the negatively charged lipids (Hu et al. 2005; Vernier et al. 2006b; Gurtovenko et al. 2010; Sridhara and Joshi 2014), only a handful simulations were performed to model the transport of large molecules (Tarek 2005; Breton et al. 2012; Salomone et al. 2014; Casciola and Tarek 2016).

1.3.4.1 Systems Subject to nsPEFs

Using high intensity electric fields, Tarek (2005) reported the first MD simulation of the transport of a short DNA double strand, showing that the uptake occurred via an electrophoretic drag and only after electropore formation in the lipid bilayer.

More recently, corroborating experiments performed on PC lipid-based giant unilamellar vesicles (GUVs), it was shown that a single 10 ns high-voltage electric pulse can permeabilize the lipid bilayer and favors the delivery of a double-stranded siRNA (−42e, 13.89 kDa) to the vesicle (Breton et al. 2012) via an electrophoretic drag through the formed electropore. Comparing the experimental evidence (mainly fluorescence imaging) with MD simulations, the authors could show in particular that (1) following the application of an electric field, the siRNA is pushed toward the lipid head-groups forming an siRNA-phospholipids head-group complex that remains stable even when the pulse is switched off; (2) no transport is detected for electric

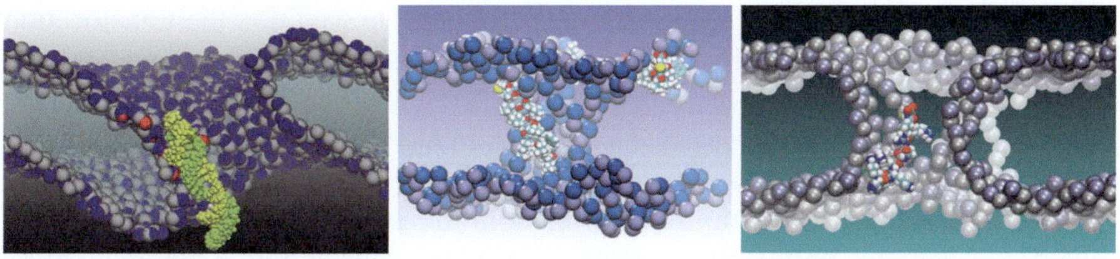

Fig. 1.2 MD simulations of the molecular interaction of small molecules with POPC membrane bilayers during the administration of ns pulsed electric fields (nsPEFs). From *left to right*: siRNA (Breton et al. 2012); CM18, residues 1–7 of cecropin-A and 2–12 of melittin antimicrobial peptides: KWKLFKKIGAVLKVLTTG (Salomone et al. 2014); and Tat11, residues 47–57 of HIV-1 Tat protein: YGRKKRRQRRR (Salomone et al. 2014). For clarity, only the phosphatidylcholine head-group atoms of the bilayer and of the transported molecules are shown

fields applied below an EP threshold; (3) when the applied E is above the EP threshold, the siRNA is electrophoretically pulled through the electropore and translocated within a 10 ns time scale (Fig. 1.2); and (4) if the electric field is turned off before the complete transition, the pore collapses around the molecule which is hence trapped.

MD simulations were also used to provide a rationale for electroporation experiments using chimeric peptides CM18-Tat_{11} (CM18, residues 1–7 of cecropin-A and 2–12 of melittin antimicrobial peptides, KWKLFKKIGAVLKVLTTG; Tat11, residues 47–57 of HIV-1 Tat protein, YGRKKRRQRRR) known to be efficient delivery vectors for plasmid DNA using endocytic vesicles. The CM18-Tat11 – pDNA complex is readily taken up by cells, accumulates within endocytic vesicles, and promotes their membrane destabilization followed by the release of cytoplasmic cargo (i.e., DNA). This requires, however, a high peptide concentration within vesicles, typically >10 μM. Salomone et al. (2014) showed that the use of nsPEFs substantially enhances the cytoplasmic DNA release. The authors modeled the peptide and its fragments and performed experiments on GUVs to provide molecular details about the processes taking place. They reported from MD simulations that, when subject to high electric fields, Tat_{11}, a small cationic peptide (+8e, 1.50 kDa) can translocate through an electroporated bilayer within a few nanoseconds without interacting with the phospholipid head-groups (Fig. 1.2). In contrast, the amphipathic peptide CM18, even when located near a preformed pore, remains anchored to the lipid head-groups and does not translocate during a 12 ns high electric field pulse (Fig. 1.2). Such a complex process, unraveled by atomistic modeling, represents a paradigmatic example that can open the way to other targeted delivery protocols.

1.3.4.2 Systems Subject to msPEFs

Two molecules, namely, the siRNA double strand and Tat_{11}, were recently investigated by our group (Casciola and Tarek 2016) in order to compare their mechanism of electric field-mediated transport under msPEFs to the one reported using the nsPEFs (Breton et al. 2012; Salomone et al. 2014).

The simulations mimicking msPEFs experiments showed that the translocation of siRNA through the pore driven by the application of TM voltages above 500 mV takes place in the 10 s of nanoseconds time scale, as reported for the nsPEFs. Notably, in both simulations carried out under an electric field or under the charge imbalance, the siRNA remained anchored to the lower leaflet of the membrane after translocation without diffusing in the bulk solution even if the voltage was maintained. Experiments performed on mouse melanoma cells applying ms-long pulses evidenced that tuning the duration of the pulse is essential for an efficient siRNA uptake. Considering the picture emerging from simulations, one can speculate that longer pulses could facilitate the transport of siRNA by the formation of a pore population having larger diameters. This would allow siRNAs to flow through the wide pore and to access the cytoplasm directly and thus increase the transport efficiency.

The translocation for Tat_{11} differs from that of the siRNA because, during the process, this peptide does not interact with lipid head-groups but rather diffuses via electrophoresis through the pore resulting in its faster uptake. Under a TM voltage of ~70 mV, the molecule is translocated from extra- to intracellular regions over the same time scale of the nsPEFs procedure (Salomone et al. 2014). At lower voltages (~ 40 mV) Tat_{11} translocates in ~30 ns, possibly a consequence of a higher hindrance of the pore (radius decreases from 2 to 0.4 nm) and of a reduction of the electrophoretic drag. Interestingly, the simulations indicate that the electrically driven uptake of a small charged molecule such as Tat_{11} through an electroporated lipid bilayer occurs on the tens of nanosecond time scale both under msPEFs and ns PEFs.

1.4 Summary and Future Directions

Atomistic simulations provide novel and needed molecular scale insight into membrane electroporation processes, thereby serving as a complementary source of information in addition to the current arsenal of experimental tools. Over the last two decades during which MD simulations have been used to decipher key aspects of the phenomena, a great and constant improvement of the protocols used has allowed better modeling of the experimental protocols, and satisfactory agreements are now being reached when comparing simulation data to "wet" lab experiments performed on similar model cell membranes (namely, liposomes). There are quite a few challenges ahead; perhaps the most important one is to develop more complex and realistic models, i.e., including a representative variety of lipids, carbohydrates, and cytoskeleton components. Furthermore, it is clear that studies so far have mainly focused on the physical changes membranes undergo when subject to PEFs, while it was reported many years ago that lipids under such conditions might also undergo chemical modification, in particular peroxidation (Benov et al. 1994; Maccarrone et al. 1995; Schnitzer et al. 2007; Zhou et al. 2007). Studying such processes would require the use of other computational chemistry modeling techniques relying rather on quantum mechanics. Though very costly, the use of such simulations would enable a full understanding and characterization of the complex phenomena of cell membrane electropermeabilization.

Acknowledgments Simulations were performed using HPC resources from GENCI-CINES. M.T. acknowledges the support of the French Agence Nationale de la Recherche, under grant (ANR-10_BLAN-916-03-INTCELL) and support from the "Contrat État Plan Region Lorraine 2015-2020" subproject MatDS. The studies were conducted in the scope of the European Associated Laboratory for Pulsed Electric Field Applications in Biology and Medicine (LEA EBAM).

References

Abidor IG, Arakelyan VB, Chernomordik LV, Chizmadzhev YA, Pastushenko VF, Tarasevich MP (1979) Electric breakdown of bilayer lipid membranes. I. The main experimental facts and their qualitative discussion. J Electroanal Chem 104:37–52

Allen MP, Tildesley DJ (1987) Computer simulation of liquids. Clarendon Press, Oxford

Anézo C, de Vries AH, Höltje HD, Tieleman DP, Marrink SJ (2003) Methodological issues in lipid bilayer simulations. J Phys Chem B 107:9424–9433

Benov LC, Antonov PA, Ribarov SR (1994) Oxidative damage of the membrane lipids after electroporation. Gen Physiol Biophys 13:85–97

Benz R, Beckers F, Zimmerman U (1979) Reversible electrical breakdown of lipid bilayer membranes – charge-pulse relaxation study. J Membr Biol 48:181–204

Berkowitz ML, Raghavan MJ (1991) Computer simulation of a water/membrane interface. Langmuir 7:1042–1044

Böckmann RA, de Groot BL, Kakorin S, Neumann E, Grubmüller H (2008) Kinetics, statistics, and energetics of lipid membrane electroporation studied by molecular dynamics simulations. Biophys J 95:1837–1850

Bowman AM, Nesin OM, Pakhomova ON, Pakhomov AG (2010) Analysis of plasma membrane integrity by fluorescent detection of Tl(+) uptake. J Membr Biol 236:15–26

Breton M, Mir LM (2012) Microsecond and nanosecond electric pulses in cancer treatments. Bioelectromagnetics 33:106–123

Breton M, Delemotte L, Silve A, Mir LM, Tarek M (2012) Transport of siRNA through lipid membranes driven by nanosecond electric pulses: an experimental and computational study. J Am Chem Soc 134:13938–13941

Cadossi R, Ronchetti M, Cadossi M (2014) Locally enhanced chemotherapy by electroporation: clinical experiences and perspective of use of electrochemotherapy. Future Oncol 10:877–890

Cascales JJL, de la Torre JG, Marrink SJ, HJC B (1996) Molecular dynamics simulation of a charged biological membrane. J Chem Phys 104:2713–2720

Casciola M, Tarek M (2016) A molecular insight into the electro-transfer of small molecules through electropores driven by electric fields. Biochim Biophys Acta Biomembr 1858:2278–2289

Casciola M, Bonhenry D, Liberti M, Apollonio F, Tarek M (2014) A molecular dynamic study of cholesterol rich lipid membranes: comparison of electroporation protocols. Bioelectrochemistry 100:11–17

Casciola M, Kasimova MA, Rems L, Zullino S, Apollonio F, Tarek M (2016) Properties of lipid electropores I: molecular dynamics simulations of stabilized pores by constant charge imbalance properties of lipid electropores I: molecular dynamics simulations of stabilized pores by constant charge imbalance. Bioelectrochemistry 109:108–116

Case DA, Darden TA, Cheatham TE III, Simmerling CL, Wang J, Duke RE, Luo R, Crowley M, Walker RC, Zhang W, Merz KM, Wang B, Hayik S, Roitberg A, Seabra G, Kolossváry I, Wong KF, Paesani F, Vanicek J, Wu X, Brozell SR, Steinbrecher T, Gohlke H, Yang L, Tan C, Mongan J, Hornak V, Cui G, Mathews DH, Seetin MG, Sagui C, Babin V, Kollman PA (2008) Amber 10. University of California, San Fransisco

Chen C, Smye SW, Robinson MP, Evans JA (2006) Membrane electroporation theories: a review. Med Biol Eng Comput 44:5–14

Chipot C, Klein ML, Tarek M, Yip S (2005) Modeling lipid membranes. In: Yip S (ed) Handbook of materials modeling. Springer, Dordrecht, pp 929–958

Chiu SW, Clark M, Jakobsson E, Subramaniam S, Scott HL (1999) Optimization of hydrocarbon chain interaction parameters: application to the simulation of fluid phase lipid bilayers. J Phys Chem B 103:6323–6327

Chiu SW, Vasudevan S, Jakobsson E, Mashl RJ, Scott HL (2003) Structure of sphingomyelin bilayers: a simulation study. Biophys J 85:3624–3635

Chopinet L, Rols M-P (2015) Nanosecond electric pulses: a mini-review of the present state of the art. Bioelectrochemistry 103:2–6

Dahlberg M, Maliniak A (2008) Molecular dynamics simulations of cardiolipin bilayers. J Phys Chem B 112:11655–11663

Damodaran KV, Merz KM (1994) A comparison of dmpc and dlpe based lipid bilayers. Biophys J 66:1076–1087

Dehez F, Delemotte L, Kramar P, Miklavčič D, Tarek M (2014) Evidence of conducting hydrophobic nanopores across membranes in response to an electric field. J Phys Chem C 118:6752–6757

Delemotte L, Tarek M (2012) Molecular dynamics simulations of lipid membrane electroporation. J Membr Biol 245:531–543

Delemotte L, Dehez F, Treptow W, Tarek M (2008) Modeling membranes under a transmembrane potential. J Phys Chem B 112:5547–5550

Deng J, Schoenbach KH, Stephen Buescher E, Hair PS, Fox PM, Beebe SJ (2003) The effects of intense submicrosecond electrical pulses on cells. Biophys J 84:2709–2714

Feller SE, Gawrisch K, MacKerell AD (2002) Polyunsaturated fatty acids in lipid bilayers: intrinsic and environmental contributions to their unique physical properties. J Am Chem Soc 124:318–326

Fernández ML, Risk M, Reigada R, Vernier PT (2012) Size-controlled nanopores in lipid membranes with stabilizing electric fields. Biochem Biophys Res Commun 423:325–330

Gennis RB (1989) Biomembranes: molecular structure and function. Springer, Heidelberg

Gurtovenko AA, Lyulina AS (2014) Electroporation of asymmetric phospholipid membranes. J Phys Chem B 118:9909–9918

Gurtovenko AA, Vattulainen I (2007) Ion leakage through transient water pores in protein-free lipid membranes driven by transmembrane ionic charge imbalance. Biophys J 92:1878–1890

Gurtovenko AA, Vattulainen I (2008) Effect of NaCl and KCl on phosphatidylcholine and phosphatidylethanolamine lipid membranes: insight from atomic-scale simulations for understanding salt-induced effects in the plasma membrane. J Phys Chem B 112:1953–1962

Gurtovenko AA, Anwar J, Vattulainen I (2010) Defect-mediated trafficking across cell membranes: insights from in silico modeling. Chem Rev 110(10):6077–6103

Hibino M, Shigemori M, Itoh H, Nagayama K, Kinosita K Jr (1991) Membrane conductance of an electroporated cell analyzed by submicrosecond imaging of transmembrane potential. Biophys J 59:209–220

Ho MC, Casciola M, Levine ZA, Vernier PT (2013) Molecular dynamics simulations of ion conductance in field-stabilized nanoscale lipid electropores. J Phys Chem B 117:11633–11640

Hu Q, Viswanadham S, Joshi RP, Schoenbach KH, Beebe SJ, Blackmore PF (2005) Simulations of transient membrane behavior in cells subjected to a high-intensity ultrashort electric pulse. Phys Rev E Stat Nonlin Soft Matter Phys 71:31914

Kakorin S, Brinkmann U, Neumann E (2005) Cholesterol reduces membrane electroporation and electric deformation of small bilayer vesicles. Biophys Chem 117:155–171

Kalinowski S, Ibron G, Bryl K, Figaszewski Z (1998) Chronopotentiometric studies of electroporation of bilayer lipid membranes. Biochim Biophys Acta 1369:204–212

Klauda JB, Venable RM, Freites JA, O'Connor JW, Tobias DJ, Mondragon-Ramirez C, Vorobyov I, MacKerell JAD, Pastor RW (2010) Update of the CHARMM all-atom additive force field for lipids: validation on six lipid types. J Phys Chem B 114:7830–7843

Koronkiewicz S, Kalinowski S (2004) Influence of cholesterol on electroporation of bilayer lipid membranes: chronopotentiometric studies. Biochim Biophys Acta Biomembr 1661:196–203

Koronkiewicz S, Kalinowski S, Bryl K (2002) Programmable chronopotentiometry as a tool for the study of electroporation and resealing of pores in bilayer lipid membranes. Biochim Biophys Acta 1561:222–229

Kotnik T, Miklavcic D (2006) Theoretical evaluation of voltage inducement on internal membranes of biological cells exposed to electric fields. Biophys J 90:480–491

Kotnik T, Miklavcic D, Slivnik T (1998) Time course of transmembrane voltage induced by time-varying electric fields – a method for theoretical analysis and its application. Bioelectrochem Bioenerg 45:3–16

Kotnik T, Pucihar G, Miklavčič D (2010) Induced transmembrane voltage and its correlation with electroporation-mediated molecular transport. J Membr Biol 236:3–13

Kotulska M (2007) Natural fluctuations of an electropore show fractional Lévy stable motion. Biophys J 92:2412–2421

Kramar P, Delemotte L, Maček Lebar A, Kotulska M, Tarek M, Miklavčič D (2012) Molecular level characterization of lipid membranes electroporation using linearly rising current experiments. J Membr Biol 245:651–659

Krassen H, Pliquett U, Neumann E (2007) Nonlinear current–voltage relationship of the plasma membrane of single CHO cells. Bioelectrochemistry 70:71–77

Kutzner C, Grubmüller H, de Groot BL, Zachariae U (2011) Computational electrophysiology: the molecular dynamics of ion channel permeation and selectivity in atomistic detail. Biophys J 101:809–817

Lakshmanan S, Gupta GK, Avci P, Chandran R, Sadasivam M, Jorge AES, Hamblin MR (2014) Physical energy for drug delivery; poration, concentration and activation. Adv Drug Deliv Rev 71:98–114

Leach AR (2001) Molecular modelling: principles and applications, 2nd edn. Prentice Hall, Harlow/Munich

Leontiadou H, Mark AE, Marrink S-J (2007) Ion transport across transmembrane pores. Biophys J 92:4209–4215

Levine ZA, Vernier PT (2010) Life cycle of an electropore: field-dependent and field-independent steps in pore creation and annihilation. J Membr Biol 236:27–36

Levine ZA, Vernier PT (2012) Calcium and phosphatidylserine inhibit lipid electropore formation and reduce pore lifetime. J Membr Biol 245:599–610

Li Z, Venable RM, Rogers LA, Murray D, Pastor RW (2009) Molecular dynamics simulations of PIP2 and PIP3 in lipid bilayers: determination of ring orientation, and the effects of surface roughness on a poisson-boltzmann description. Biophys J 97:155–163

Maccarrone M, Rosato N, Agrò AF (1995) Electroporation enhances cell membrane peroxidation and luminescence. Biochem Biophys Res Commun 206:238–245

Miklavčič D, Mali B, Kos B, Heller R, Serša G (2014) Electrochemotherapy: from the drawing board into medical practice. Biomed Eng Online 13:29

Mir LM, Banoun H, Paoletti C (1988) Introduction of definite amounts of nonpermeant molecules into living cells after electropermeabilization – direct access to the cytosol. Exp Cell Res 175:15–25

Mukhopadhyay P, Monticelli L, Tieleman DP (2004) Molecular dynamics simulation of a palmitoyl-oleoyl phosphatidylserine bilayer with Na+ Counterions and NaCl. Biophys J 86:1601–1609

Nesin OM, Pakhomova ON, Xiao S, Pakhomov AG (2011) Manipulation of cell volume and membrane pore comparison following single cell permeabilization with 60- and 600-ns electric pulses. Biochim Biophys Acta Biomembr 1808:792–801

Neumann E, Sowers AE, Jordan CA (1989) Electroporation and electrofusion in cell biology. Plenum Press, New York

Pakhomov AG, Gianulis E, Vernier PT, Semenov I, Xiao S, Pakhomova ON (2015) Multiple nanosecond electric pulses increase the number but not the size of long-lived nanopores in the cell membrane. Biochim Biophys Acta 1848:958–966

Pandit AS, Bostick D, Berkowitz ML (2003) Molecular dynamics simulation of a dipalmitoylphosphatidylcholine bilayer with NaCl. Biophys J 84:3743–3750

Patel RY, Balaji PV (2008) Characterization of symmetric and asymmetric lipid bilayers composed of varying concentrations of ganglioside GM1 and DPPC. J Phys Chem B 112:3346–3356

Pauly H, Schwan HP (1959) Uber Die Impedanz Einer Suspension Von Kugelformigen Teilchen Mit Einer Schale – Ein Modell Fur Das Dielektrische Verhalten Von Zellsuspensionen Und Von Proteinlosungen. Z Naturforsch B 14:125–131

Pavlin M, Kandušer M, Reberšek M, Pucihar G, Hart FX, Magjarevićcacute, Ratko, Miklavčič D (2005) Effect of cell electroporation on the conductivity of a cell suspension. Biophys J 88:4378–4390

Piggot TJ, Holdbrook DA, Khalid S (2011) Electroporation of the E. coli and S. aureus membranes: molecular dynamics simulations of complex bacterial membranes. J Phys Chem B 115:13381–13388

Polak A, Bonhenry D, Dehez F, Kramar P, Miklavčič D, Tarek M (2013) On the electroporation thresholds of lipid bilayers: molecular dynamics simulation investigations. J Membr Biol 246:843–850

Polak A, Tarek M, Tomšič M, Valant J, Ulrih NP, Jamnik A, Kramar P, Miklavčič D (2014) Electroporation of archaeal lipid membranes using MD simulations. Bioelectrochemistry 100:18–26

Pucihar G, Kotnik T, Miklavcic D, Teissie J (2008) Kinetics of transmembrane transport of small molecules into electropermeabilized cells. Biophys J 95:2837–2848

Rög T, Murzyn K, Pasenkiewicz-Gierula M (2002) The dynamics of water at the phospholipid bilayer: a molecular dynamics study. Chem Phys Lett 352:323–327

Rog T, Martinez-Seara H, Munck N, Oresic M, Karttunen M, Vattulainen I (2009) Role of cardiolipins in the inner mitochondrial membrane: insight gained through atom-scale simulations. J Phys Chem B 113:3413–3422

Roux B (1997) Influence of the membrane potential on the free energy of an intrinsic protein. Biophys J 73:2980–2989

Sachs JN, Crozier PS, Woolf TB (2004) Atomistic simulations of biologically realistic transmembrane potential gradients. J Chem Phys 121:10847–10851

Saiz L, Klein ML (2002) Computer simulation studies of model biological membranes. Acc Chem Res 35:482–489

Salomone F, Breton M, Leray I, Cardarelli F, Boccardi C, Bonhenry D, Tarek M, Mir LM, Beltram F (2014) High-yield nontoxic gene transfer through conjugation of the CM 18-Tat11 chimeric peptide with nanosecond electric pulses. Mol Pharm 11:2466–2474

Schnitzer E, Pinchuk I, Lichtenberg D (2007) Peroxidation of liposomal lipids. Eur Biophys J 36:499–515

Schuler LD, Daura X, van Gunsteren WF (2001) An improved GROMOS96 force field for aliphatic hydrocarbons in the condensed phase. J Comput Chem 22:1205–1218

Silve A, Leray I, Mir LM (2012) Demonstration of cell membrane permeabilization to medium-sized molecules caused by a single 10 ns electric pulse. Bioelectrochemistry 87:260–264

Sridhara V, Joshi RP (2014) Numerical study of lipid translocation driven by nanoporation due to multiple high-intensity, ultrashort electrical pulses. Biochim Biophys Acta Biomembr 1838:902–909

Szabo M, Wallace MI (2015) Imaging potassium-flux through individual electropores in droplet interface bilayers. Biochim Biophys Acta 1858(3):613–617

Tarek M (2005) Membrane electroporation: a molecular dynamics simulation. Biophys J 88:4045–4053

Tarek M, Delemotte L (2010) Electroporation of lipid membranes. In: Advanced Electroporation Techniques in Biology and Medicine. CRC Press, Boca Raton, pp 141–160

Tarek M, Tobias DJ, Chen SH, Klein ML (2001) Short wavelength collective dynamics in phospholipid bilayers: a molecular dynamics study. Phys Rev Lett 87:238101

Teissie J (2013) Electrically mediated gene delivery: Basic and translational concepts, novel gene therapy approaches, Prof. Ming Wei (ed.), InTech, doi:10.5772/54780. Available from: https://www.intechopen.com/books/novel-gene-therapy-approaches/electrically-mediated-gene-delivery-basicand-translational-concepts

Teissié J, Eynard N, Gabriel B, Rols MP (1999) Electropermeabilization of cell membranes. Adv Drug Deliv Rev 35:3–19
Tieleman DP (2004) The molecular basis of electroporation. BMC Biochem 5:10
Tobias DJ, Tu K, Klein ML (1997) Assessment of all–atom potentials for modeling membranes: molecular dynamics simulations of solid and liquid alkanes and crystals of phospholipid fragments. J Chim Phys 94:1482–1502
Vacha R, Berkowitz ML, Jungwirth P (2009) Molecular model of a cell plasma membrane with an asymmetric multicomponent composition: water permeation and ion effects. Biophys J 96:4493–4501
Vernier PT, Ziegler MJ (2007) Nanosecond field alignment of head group and water dipoles in electroporating phospholipid bilayers. J Phys Chem B 111:12993–12996
Vernier PT, Sun Y, Gundersen MA (2006a) Nanoelectropulse-driven membrane perturbation and small molecule permeabilization. BMC Cell Biol 7:37
Vernier PT, Ziegler MJ, Sun Y, Gundersen MA, Tieleman DP (2006b) Nanopore- facilitated, voltage-driven phosphatidylserine translocation in lipid bilayers- in cells and in silico. Phys Biol 3:233–247
Villemejane J, Mir LM (2009) Physical methods of nucleic acid transfer: general concepts and applications. Br J Pharmacol 157:207–219
Weaver JC (2003) Electroporation of biological membranes from multicellular to nano scales. IEEE Trans Dielectr Electr Insul 10:754–768
Weaver JC, Chizmadzhev YA (1996) Theory of electroporation: a review. Bioelectrochem Bioenerg 41:135–160
Yarmush ML, Golberg A, Serša G, Kotnik T, Miklavčič D (2014) Electroporation-based technologies for medicine: principles, applications, and challenges. Annu Rev Biomed Eng 16(1):295
Zhou Y, Berry CK, Storer PA, Raphael RM (2007) Peroxidation of polyunsaturated phosphatidylcholine lipids during electroformation. Biomaterials 28:1298–1306
Ziegler MJ, Vernier PT (2008) Interface water dynamics and porating electric fields for phospholipid bilayers. J Phys Chem B 112:13588–13596

Chapter 2
Role of Bioinformatics in the Study of Ionic Channels

Monika Kurczyńska, Bogumił M. Konopka, and Małgorzata Kotulska

Abstract Ionic channels belong to the group of the most important proteins. Not only do they enable transmembrane transport but they are also the key factors for proper cell function. Mutations changing their structure and functionality often lead to severe diseases called channelopathies. On the other hand, transmembrane channels are very difficult objects for experimental studies. Only 2% of experimentally identified structures are transmembrane proteins, while genomic studies show that transmembrane proteins make up 30% of all coded proteins. This gap could be diminished by bioinformatical methods which enable modeling unknown protein structures, functions, transmembrane location, and ligand binding. Several in silico methods dedicated to transmembrane proteins have been developed; some general methods could also be used. They provide the information unavailable from experiments. Current modeling tools use a variety of computational methods, which provide results of surprisingly high quality.

2.1 Introduction

Ionic channels belong to the group of transmembrane proteins, spanning membranes of each living cell. They are responsible for the passive transport of ions exchanged between cytoplasm and extracellular environment, as well as between various cell compartments. Ionic channels play important roles in the human organism: regulation and control of ionic transport from and to the cell. They participate in nerve and

M. Kurczyńska • B.M. Konopka • M. Kotulska (✉)
Department of Biomedical Engineering, Faculty of Fundamental Problems of Technology, Wroclaw University of Science and Technology, 50-370 Wroclaw, Poland
e-mail: malgorzata.kotulska@pwr.edu.pl

J. Kulbacka, S. Satkauskas (eds.), *Transport Across Natural and Modified Biological Membranes and its Implications in Physiology and Therapy*, Advances in Anatomy, Embryology and Cell Biology 227,
DOI 10.1007/978-3-319-56895-9_2

muscle impulse transfer, hormone regulation, blood pressure regulation, and water regulation (Restrepo-Angulo et al. 2010). Channels are involved in all major processes and constitute an important target for many pharmaceuticals. Their proper functioning, which can be severely disturbed by mutations, is crucial to maintain health. Unfortunately, ionic channels are molecules that are very difficult to study. They are usually large proteins with sequences up to even a few thousands of amino acids. Moreover, preparation of a protein for structural studies may be problematic (Koehler Leman et al. 2015). The host cells used for the protein expression often cause its over- or under-expression, as well as misfolding. Obtaining a high-resolution structure of a membrane protein often requires its removal from the natural lipid environment by denaturation, which may affect its native conformation. Membrane lipids and proteins mutually influence their structures, so the ideal case for structural studies requires membrane-mimetic detergents or lipid media such as lipidic cubic phase (LCP) (Landau and Rosenbusch 1996) or lipidic medium for crystallization-bicelles (Ujwal and Abramson 2012). Continuous development of the experimental methods, such as NMR methodologies for membrane protein, oriented sample NMR and magic angle spinning (MAS) in solid-state NMR (ssNMR) spectroscopy (Wanga and Ladizhansky 2014), has contributed to doubling the growth of solved membrane protein structures for the last 3 years. However, difficulties in experimental efforts to clarify membrane protein structures and functions remain. Therefore, studies in silico could provide a valuable support to obtain new knowledge. Computational methods can be based either on results of other experiments, gathered in molecular databases, or a complete de novo modeling using available services and tools or novel algorithms developed by researchers.

Accessible online services in molecular biology provide information about biological sequences, molecular structures, congenital diseases, and their molecular mechanisms. They include databases, knowledge bases, search, modeling, and evaluation tools. Therefore, they are crucial for today's researchers as well as for physicians. Molecular Biology Databases Collections have been published in the journal of Nucleic Acids Research, in the first released issue of each year for 24 years. These issues gather information about all "bio" databases available at the time. Currently the number of databases has reached 1712, and they are divided into 14 categories (Galperin et al. 2017).

The knowledge of a protein is much more useful if its sequence is complemented with a protein three-dimensional structure. Only then could it be effectively applied for diagnosis, treatment, and preventing diseases. The first solved protein structures, myoglobin and hemoglobin, were published in 1958 and 1963, respectively (Kendrew et al. 1958; Muirhead and Perutz 1963), but their resolutions were insufficient to create models with all atomic details. The first high-resolution structure of the hen egg-white lysozyme was published in 1965 (Blake et al. 1965). However, the first high-resolution structure of an ionic channel, a potassium channel from *Streptomyces lividans* was only solved in 1998 by the group of R. McKinnon (Doyle et al. 1998), who has been awarded with the Nobel Prize for this discovery. Now about 120,000 protein structures are known and collected in Protein Data Bank (PDB) (Berman et al. 2000). Only 2% of them are membrane proteins although the genetic studies imply that they should constitute around 30%. Ionic channels are represented by only 137 structures, and almost none of them are from a human tissue. Most of the experimental membrane protein structures have been obtained with X-ray crystallography (3651). Other methods include NMR spectroscopy, electron microscopy, hybrid methods, electron crystallography, neutron diffraction, and fiber diffraction.

The low number of the membrane protein structures requires optimized tools and databases to take full advantage of the knowledge collected so far. Some databases, such as Membrane Proteins of Known 3D Structure (http://blanco.biomol.uci.edu/mpstruc/), gather information about membrane protein structures in one place, and they are linked to the PDB database or different sources of additional knowledge about them. There are some databases of 3D structures which include information about location of the membrane proteins embedded in lipid bilayers, e.g., PDBTM (Kozma et al. 2013) (http://pdbtm.enzim.hu/), OPM (Lomize et al. 2012) (http://opm.phar.umich.edu/), and MemProtMD (Stansfeld et al. 2015) (http://sbcb.bioch.ox.ac.uk/memprotmd/beta/). Other databases are dedicated to special cases of membrane proteins, for example, Membranome (Lomize et al. 2016) (http://membranome.org/), which collects single-spanning (bitopic) transmembrane proteins from six organisms, or TCDB (Saier et al. 2016) http://tcdb.org/), which includes functional and phylogenetic classification of membrane transport proteins. Many databases gather information about topology and sequence motifs of membrane proteins, e.g., TOPDB (Tusnády et al. 2008a) (http://topdb.enzim.hu/), ExTopoDB (Tsaousis et al. 2010) (http://bioinformatics.biol.uoa.gr/ExTopoDB/), TOPDOM (Tusnády et al. 2008b) (http://topdom.enzim.hu/), and MeMotif (Marsico et al. 2010). Some databases collect information about genetic variations in the genes encoding ionic channels and resulting diseases, such as the Ion Channels Variants Portal (Hinard et al. 2017) (https://www.nextprot.org/portals/navmut).

A great interest in the membrane protein structures and functions versus insufficient structural information from experimental methods are the reasons of a wide use of bioinformatics methodology and tools to explore membrane proteins. Bioinformatics may not only support high-resolution prediction or refinement of membrane protein structures but also help in predicting the effects of mutations, ligand docking, drug design, and loop modeling. Other challenges involve prediction of the protein functionality, such as ion selectivity mechanisms, channel gating, sensory functions, and ligand binding.

In the next two sections of the chapter, we will review various tools and methods used in bioinformatical modeling of the structure and function of ionic channels. This will be followed by a short review of applications of those methods in studying diseases related to ionic channel dysfunctions.

2.2 Modeling of Ionic Channels

2.2.1 Structure Prediction

Typically, modeling of a protein structure is carried out on several levels with the ultimate goal being generation of a reliable three-dimensional model. These levels include prediction of the secondary structure, disulfide bonds, solvent accessibility, and inter-residue contacts. This information, along with the amino acid sequence, is used as an input for tools predicting the 3D models. Usually, the last step of 3D prediction protocols is the model quality assessment. Most methods that perform the abovementioned tasks are of general use. Thus, they can be applied to any type of protein including membrane proteins. However, there are also issues specific to the transmembrane protein structure prediction.

Topology prediction of transmembrane (TM) proteins is the first specific task to be performed in the procedure of predicting structures of ionic channels. Tools for predicting TM topology fall into two main categories: helical TM topology predictors and beta-barrel TM topology predictors.

Helical TM topology can be predicted from the first principles. SCAMPI and TopPred$^{\Delta G}$ (Bernsel et al. 2008) are two methods that predict the TM topology based on an experimentally derived model, which estimates the free energy of insertion of an amino acid into the membrane. The difference is that SCAMPI additionally uses a simple model similar to hidden Markov model (HMM) that defines the underlying grammar, while TopPred$^{\Delta G}$ uses a sliding window to generate a free energy profile and data-based energy cutoffs to mark "certain" and "putative" TM segments.

However, most helical TM topology predictors are machine-learning methods. MemBrain (Shen and Chou 2008) uses a multiple sequence alignment-based position scoring matrix, which is cut into 13 and 15 amino acid long frames with a sliding window. The windows are input to k-nearest neighbor classifiers to generate TM helix propensity profiles that are used to identify TM regions. MEMSAT-SVM (Nugent and Jones 2009) uses four SVM binary classifiers that are used to identify the following regions: inside/outside loops, TM helix/non-TM helix, reentrant helix/non-reentrant helix, and signal peptide/non-signal peptide. The input is the position scoring matrix generated along with a multiple sequence alignment. TMHMM (Krogh et al. 2001) is a method that implements a hidden Markov model of a transmembrane helical protein. The model assumes three regions of a TM protein: extracellular, transmembrane, and cytoplasmic sides. Each region is modeled by a submodel that contains several HMM states. HMMTOP (Tusnady and Simon 2001), similarly to TMHMM, applies HMM. The model enables recognizing parts of a sequence and assigns them into one of five regions: inside loop, inside helix tail, membrane helix, outside helix tail, and outside loop. Other HMM-based methods include, for instance, PRO-TMHMM, PRODIV-TMHMM (Viklund and Elofsson 2004), and Phobius (Kall et al. 2007). More elaborate TM predictors use mixed approaches. For instance, OCTOPUS (Viklund and Elofsson 2008) uses PSI-BLAST position-specific scoring matrices (PSSM) and raw sequence alignments as inputs for a two-stage modeling procedure. First, a set of artificial neural networks is used to predict residue propensities to be a part of the membrane, interface, loop, or globular regions. The output is then fed into an HMM model that is used to derive the final topology prediction.

Quite successful are consensus approaches which use outputs from several standalone methods for calculating a consensus prediction. TOPCONS (Bernsel et al. 2009) is based on the outputs from five TM topology prediction methods. These outputs are combined into a topology profile which is further used as the input to a dynamic programming algorithm that generates a final prediction. CCTOP (Dobson et al. 2015) uses the output of ten TM topology predictors as weighted constrains that are fed into an HMM model. In addition, CCTOP may utilize other available data, such as conservatively localized motifs and domains.

In an assessment of secondary structure predictors, which we performed several years ago, we showed that prediction of beta structures is much harder than prediction of alpha helices (Konopka et al. 2009). The works on prediction of beta-barrel TM topology prediction published since then clearly show the significant progress in this area.

PRED-TMBB (Bagos et al. 2004) is one of the first methods designed to model the topology of beta-barrel TM proteins. It was able to predict the topology of outer membrane beta-barrel TM proteins of gram-negative bacteria using an HMM model. In the paper of Taylor et al. (2006), a Bayesian network-based method predicting beta-barrel TM topology was proposed. The authors reported high prediction accuracy of a beta strand (88.6%); however, the accuracy in predicting the overall topology was quite low (42%). PROFtmb (Bigelow and Rost 2006) implements a HMM-based model that allows four-state residue prediction. The regions include up-strand, down-strand, periplasmic loop, and outer loop. The reported prediction accuracy, calculated per residue, achieved 86%. BOCTOPUS (Hayat and Elofsson 2012) and its successor BOCTOPUS2 (Hayat et al. 2016) are among the most successful beta-barrel TM topology predictors. They use a combination of SVMs and HMM to predict the topology of TM beta-barrel proteins. To be more specific, BOCTOPUS2 is a three-stage predictor. It uses a position-specific scoring matrix (PSSM) as input information for SVM classifiers that assess the probability of each residue belonging to one of four β-strand regions, i.e., outer loop (o), inner loop (i), pore-facing (p), or lipid-facing (l). The barrel region is localized according to the result. Finally, an HMM-like model is used to predict the topology. PredβTM (Roy Choudhury and Novic 2015) uses an interesting input sequence coding scheme, according to which the sequence is divided into ten amino acid long segments. For each segment, an amino acid adjacency matrix is created, which is then fed into a SVM classifier. The method yields accuracies worse than that of BOCTOPUS2.

The TM topology prediction methods that are available as web servers are listed in Table 2.1.

Topology prediction can be used as input information that supports 3D modeling. In the best-case prediction scenario, when structures of homologues of the target structure are available, TM topology is not as important as the quality of templates. In these cases, homology modeling methods are used. Homology modeling includes three important steps: (1) template search/fold recognition, (2) alignment, and (3) coordinate generation with loop modeling.

When performing homology modeling of TM proteins, one can use general-purpose modeling tools, such as SWISS-MODEL (Biasini et al. 2014), Phyre2 (Kelley et al. 2015), HHpred (Soding et al. 2005), and MODELLER (Webb and Sali 2016). However, tools which were specially tailored to TM protein homology modeling could also be applied. MEDELLER (Kelm et al. 2010) is probably one of the

Table 2.1 TM topology predictors available as web server services

Method name	Methodology	Web server address
HMMTOP2.0	Single sequence, HMM-based model	http://www.enzim.hu/hmmtop
TMHMM2.0	Single sequence, HMM-based model	http://www.cbs.dtu.dk/services/TMHMM/
Phobius	Single sequence, HMM-based model	http://phobius.binf.ku.dk/
MEMSAT-SVM	MSA-based, SVM predictor	http://bioinf.cs.ucl.ac.uk/psipred/
MEMSAT3	MSA-based, neural network predictor	http://bioinf.cs.ucl.ac.uk/psipred/
MemBrain	MSA-based k-nearest neighbor predictor	http://www.csbio.sjtu.edu.cn/bioinf/MemBrain/
OCTOPUS	MSA-based neural network predictor combined with HMM	http://octopus.cbr.su.se/
SCAMPI-msa	Physics-based model	http://scampi.cbr.su.se/
TOPCONS	Consensus predictor	http://topcons.net/
CCTOP	Consensus predictor	http://cctop.enzim.ttk.mta.hu
Beta-barrel TM proteins		
PRED-TMBB	Single sequence, HMM-based model	http://bioinformatics.biol.uoa.gr/PRED-TMBB/
BOCTOPUS2	MSA-based, SVM combined with a HMM	http://boctopus.bioinfo.se/
PROFtmb	MSA-based, HMM model	https://www.predictprotein.org/
PredβTM	Single sequence, SVM predictor	http://transpred.ki.si/

The *top part* of the table lists helical TM predictors, and the *lower part* lists beta-barrel TM predictors

most popular tools. It is based on MODELLER as the main atom coordinate generation engine; however, it uses additional information on membrane positioning to identify the most reliable part of the template for the target alignment. According to its authors, it consistently yields better results for membrane proteins than MODELLER alone. Another interesting approach is presented in Chen et al. (2014). In this work, a generic protocol for modeling TM helical proteins is provided. It includes standard homology modeling using MEDELLER in cases where a high-quality alignment to structural templates is available, while in cases where alignments contain significant gaps, additional modeling steps are performed, i.e., low-energy TM helix conformation search with interhelical bends and kink insertion and modeling the regions around them using techniques of fragment insertion.

There are several fold recognition and alignment tools that may improve the result of homology modeling by providing better structural templates and better alignments than general tools. TMFR (Wang et al. 2013) is a fold recognition tool designed to search TM proteins' homologues. It benefits from inclusion of TM protein-specific topology features, which are not used by other general fold recognition tools. Apart from finding homology modeling templates, it can also be used to improve the alignment between the target and the template structure. PRALINETM (Pirovano et al. 2008) is a multiple alignment method that uses membrane-specific substitution matrix in those regions of the alignment that were annotated as TM regions by topology predictor methods. MT-T (Hill and Deane 2013) is a similar approach. It also uses environment-specific substitution tables. Different scoring and gap penalty scheme is used depending on whether the analyzed region is a TM region or not. The tool, using homology modeling procedure, shows improved alignment accuracy and better final models in comparison to similar tools of general use. TM-Coffee (Chang et al. 2012) is in fact the T-Coffee alignment tool (Notredame et al. 2000); however, it uses a reduced database only limited to TM proteins, thus acquiring alignment accuracy similar to that provided by T-Coffee but at a significantly reduced computational cost.

If no homologous structures are available, de novo modeling is performed. In this case, predictions often benefit from additional information of the secondary structure, disulfide bonds, solvent accessibility, or inter-residue contacts. The latter turned out to be especially useful due to a breakthrough in methods for predicting contact sites which are based on correlated mutations. In Konopka et al. (2014) and Dyrka et al. (2016), it was shown that it is possible to acquire high-quality structures of TM proteins based solely on contact information. However, in a real-life situation, full contact information is not available. Instead, predicted contacts are used to guide the prediction process.

Among the most successful de novo tools are Rosetta, FILM3, and EVFold_bb. Rosetta is one of the first tools for protein de novo modeling (Simons et al. 1997). There is a Rosetta protocol dedicated to membrane protein structure prediction, i.e., RosettaMembrane (Yarov-Yarovoy et al. 2006; Barth et al. 2007). The protocol uses the standard Rosetta ab initio protocol, which samples the protein conformational space by a fragment assembly algorithm, guided by an energy scoring function. However, for membrane proteins, it is enhanced with a membrane-specific physical model of intra-protein and protein-solvent interactions. Similarly to Rosetta, FILM3 (Nugent and Jones 2012) is a fragment assembly approach. However, in case of FILM3, the assembly process is guided exclusively by energy potential predictions of inter-residue contacts, derived by PSICOV from correlated mutations (Jones et al. 2012). EVFold_bb (Hayat et al. 2015) is dedicated to 3D modeling of beta-barrel proteins. It combines inter-residue contacts from the EVFold_PLM (Marks et al. 2011) with beta-barrel topology predictions from BOCTOPUS2. These contacts are then used as spatial constraints for CNS structure reconstruction software (Brunger 2007).

2.2.2 *Scoring Functions and Force Fields*

In bioinformatics of ionic channels, a few terms are used to describe the energy of the system. An individual potential, which describes a particular type of interactions in the system with a mathematical function, is often called an energy term. Scoring functions, which are sums of individual energy terms, enable choosing the best protein models generated by prediction methods. In molecular simulations, the force fields are sums of individual energy terms and are used to describe the potential energy of the system. The scoring functions and molecular force field parameters for membrane proteins are the area of active research. Energy terms are usually divided into three types: physics-based, knowledge-based, and combination of the two.

Physics-based potentials typically include energy terms describing covalent bonds (described by a harmonic potential), angles (harmonic potential), dihedral angles, electrostatics (Coulomb law), and van der Waals interactions (Lennard-Jones potential). The primary physics-based potentials are known as molecular mechanics (MM) force fields. MM force fields obtain parameters from experimental data, and quantum chemistry calculations carried out on small molecules (Weiner et al. 1984). Next, the parameters are validated by comparisons to the experimental observations such as intermolecular energies in the gas phase, solution phase densities, and heats of vaporization (Jorgensen et al. 1996). However, there is no simple way to transfer MM force fields from small molecules to proteins and macromolecular systems. This challenge has been successfully accepted in the 1990s. Originally, the native structure in a metastable conformation was demonstrated with molecular dynamics (MD) (MacKerell et al. 2000), but then larger ranges of conformational space of protein structure were studied (Shaw et al. 2010). MD is based on classical Newtonian physics. The trajectories of all the atoms or superatoms (coarse-grain version) in the simulation system, including protein and water molecules, are calculated at each time step. This approach requires a great deal of computational power and in return gives the opportunity to explain experimental observations using atomic resolution. MD studies of membrane proteins may support better understanding of the phenomena such as ionic channel gating (Meng et al. 2016; Hulse et al. 2014) and transport across the membrane, including selective transports of metal ions across the membrane (Joh et al. 2014). MD simulations may include all atoms in the system as a single practice (all-atom MD) or as larger superatoms defined by small groups of atoms (coarse-grain MD). Depending on the strategies and objectives, different force fields and different types of MD simulations are chosen. Popular force fields used in MD of biomolecules include CHARMM (Mackerell et al. 1998), AMBER (Cornell et al. 1995), GROMOS (Oostenbrink et al. 2004), and OPLS (Jorgensen et al. 1996). Typical force field for lipid membrane and then extended for proteins is MARTIN (Monticelli et al. 2008). Originally, it was created for coarse-grain MD simulations. However, MARTIN has recently been combined with the GROMOS all-atom force field (Wassenaar et al. 2013), and a membrane protein is modeled in all-atom scale while lipids in coarse-grain MD. The popular force fields have been developed for membrane protein simulations: CHARMM with lipids (Klauda et al. 2010), OPLS-AA with lipid chains (Siu et al. 2012), OPLS-AA with DPPC molecules (Maciejewski et al. 2014), and GROMOS with PC, PE, and PG lipids (Chiu et al. 2009). More details about the force fields for simulations of membranes and proteins are presented in the review by Pluhackova and Bockmann (2015). However, not only a force field is required for MD simulations but also an appropriate choice of boundary conditions, pressure, and temperature coupling schemes, reviewed by Tieleman (2012). MD simulations also need topology files for lipids, which are constantly collected in Lipidbook (Domanski et al. 2010), a public repository of parameters for MD simulations of biomembrane components.

To improve the speed of simulation, a number of methods can be applied. Replica-exchange sampling (REMD, or parallel tempering) (Ulmschneider et al. 2014) creates identical replicas of a system, which are simulated in parallel using exponentially distributed temperatures. Dynamic importance sampling (DIMS) makes sampling transition states between two known protein conformations (Seyler and Beckstein 2014). For MD simulations, several software packages were created: GROMACS (Berendsen et al. 1995), NAMD (Phillips et al. 2005), Desmond (Shaw 2005), and CHARMM (Brooks et al. 1983). In recent years, a lot of ionic channel mechanisms have been explained using MD simulations. The MD simulations were used for studies of water movement in ionic channels (Aityan and Chizmadzhev 1986), ion-specific diffusion through transmembrane protein channels (Fischer and Brickmann 1983), and solvent effects in ionic transport (Kappas et al. 1985). One of the first MD simulation targets was gramicidin A (Roux and Karplus 1994; Liu et al. 2005). However, the potassium channel has been the most popular narrow ionic channel for MD simulations (Burykin et al. 2003; Monticelli et al. 2004), because it was the first ionic channel structure discovered with high resolution using X-ray crystallography. The MD simulations of potassium channels included explanation of several processes such as conformational dynamics during activation and upon ion binding, mechanism of ion selectivity (Medovoy et al. 2016), and interactions with peptides, proteins, and lipids (Molina et al. 2015).

Knowledge-based energy functions usually take parameters from macromolecular structures, such as protein structures. They are based on the assumption that structural protein properties are independent, and the distance between two interaction sites follows Boltzmann distribution (Jernigan and Bahar 1996). Unfortunately, they often ignore physical chemistry of interactions between atoms. However, they have a wide range of applications, such as scoring functions in tools for prediction of the protein structures, for example, TMFoldRec (Kozma and Tusnády 2015). The energy function for a given conformation and sequence is calculated as a sum of products of interaction energies between the *i*th and the *j*th type of amino acids and interaction energy between the *i*th type amino acid and a lipid. The interaction energies between the amino acid pairs and between the amino acids and their lipid ambient were calculated with a modified ENERGI algorithm (Thomas and Dill 1996). Another tool, BCL::MP-Fold (Weiner et al. 2013), which is for de novo prediction of membrane protein helical bundles, uses a knowledge-based score. It uses BCL::Score for soluble proteins (Woetzel et al. 2012) with two modifications, radius of gyration and amino acid environment, and introduces three new scores, SSE alignment, membrane protein topology, and environment prediction accuracy. BCL::Score is derived using Bayes' theorem and describes the probability of observing the structure for given sequence using two separated terms: the probability of observing the structure independent of the sequence and the probability of observing the sequence for given structure. TransFold (Clote et al. 2006) tool, which is a membrane protein beta-barrel predictor, also uses pairwise interstrand residue statistical potentials, which describe number of transmembrane beta-strands in the barrel, length of transmembrane beta-strands, strand inclination with respect to membrane plane, size of periplasmic and extracellular loops, and hydrophobic profile of transmembrane beta-strands. It also uses program BETAWRAP (Cowen et al. 2002) for calculating contact potentials. PartiFold (Waldispuhl et al. 2008) for membrane beta-barrel proteins uses a novel energy function of stacked amino acid pair statistical potentials.

Methods for improvement of scoring functions or force fields combine the strengths of the two types of the energy terms: physics and knowledge-based (Song et al. 2011). The combination strategy is used in TMBPro (Randall et al. 2008), which is de novo predictor of membrane protein beta-barrels. In this case, energy terms favor formation of predicted beta-contacts, compliance to a specific membrane channel pattern, some side-chain interactions between predicted non-beta positions (Zhang et al. 2003), and close termini proximity at artificial chain break sites. At the same time, they penalize clashes between side-chain centers of mass and steric clashes between all explicitly modeled atoms using van der Waals radii. One of the most successful de novo tools, which uses a combination of the physical- and knowledge-based energy terms, is Rosetta. The Rosetta energy function is a linear combination of energy terms which describe interactions between atoms modeled with Lennard-Jones terms, solvation effect (Lazaridis and Karplus 1999), orientation-dependent H-bond term (Kortemme et al. 2003), side-chain and backbone torsion energies, short-range knowledge-based electrostatic term, and reference energies for each of the amino acids. The default energy function in Rosetta is *talaris2014*, which is a slight modification of the *talaris2013* energy function (Leaver-Fay et al. 2013). Using both types of energy terms may lead to double counting of the same physical interaction, which may provide inaccurate results. However, the problem was resolved by an iterative approach to detect it, based on the properties of energy minima distributed throughout conformational space. The RosettaMP (Alford et al. 2015) software was developed specially for membrane proteins. It combines the energy terms from Rosetta, knowledge-based energy terms derived from a database of membrane protein structures, and physics-based energy terms for atoms in the lipid membrane based on the Lazaridis implicit Gaussian-exclusion solvation model (Lazaridis 2003). RosettaMP knowledge-based energy terms, dedicated for membrane, describe propensity of a single residue to be at a given depth in the membrane, and burial by residues, pairwise interaction potential between two residues being some distance apart at a given depth in the membrane, residue density potential based on the number of neighboring residues and a number of transmembrane helices, agreement between predicted lipophilicity (from LIPS server) and the model (Adamian et al. 2005), and propensity of a single atom to be at a given depth in the membrane. They also include penalties for non-helical secondary structure in the membrane, residues outside the hydrophobic layer of the membrane, and transmembrane helices that project outside of the membrane. RosettaMP physics-based energy terms dedicated for membrane define free energy of a single, isolated atom in a solvent or lipid, depending on the depth in the membrane, and atomic solvation free energy change due to the presence of surrounding atoms (Yarov-Yarovoy et al. 2006). RosettaMP also uses depth-adjusted Rosetta hydrogen bonding term with stronger hydrogen bonding in the membrane (Kortemme et al. 2003).

2.2.3 Functional Modeling

The detailed description of the transport mechanism is crucial in ionic channel studies, especially ion-binding sites, permeation pathways, ion conductance, selective permeability, and gating. Again, computational methods may be used to answer the questions related to these issues. Moreover, modeling the channel function can be used in evaluating quality of protein structures obtained from structural modeling. Improper flow characteristics may indicate errors in the structure. The most common simulation methods include MD, Brownian dynamics (BD), and the electrodiffusion Poisson-Nernst-Planck model (PNP).

MD enables obtaining explanation of ionic channel at atomic resolution. When more structures of ionic channels were discovered, MD simulations were used to study ion transport in different proteins. In PubMed, more than 1900 articles about MD simulations of ionic channels can be found, including, for example, voltage-gated sodium channels (Oakes et al. 2016), voltage-gated cation channels (VGCC) (Delemotte et al. 2012), cationic mechanosensitive channels (MSCs) (Gnanasambandam et al. 2017), and cystic fibrosis transmembrane conductance regulator (CFTR) protein of an ABC transporter that functions as an ATP-gated channel (Callebaut et al. 2017).

BD is a less detailed method of the channel flow modeling. It treats solvent as a continuous medium while ions as discrete objects. Calculations are based on the Langevin's stochastic equation (Chung and Kuyucak 2002):

$$m_i \frac{dv_i}{dt} = - m_i \gamma_i v_i + R_i + \mathrm{Fs}_i \tag{2.1}$$

where v_i is the velocity of ith ion, γ_i is the friction coefficient of ith ion, R_i is the random force, and Fs_i represents the systematic force. The random and friction forces depict the ion-water molecules collisions. There are several available tools for BD simulations, such as GCMC/BD (Lee et al. 2012), BD_BOX (Dlugosz et al. 2011), IMBD (Kurczynska and Kotulska 2014), and BROWNIES (Berti et al. 2014). The BD simulations were used in ionic channels to study ion flow through porin channels (OmpF, alpha-hemolysin) (Schirmer and Phale 1999; Im and Roux 2002), anion channels (ClC, VDAC) (Coalson and Cheng 2010; Krammer et al. 2013), cation channels (Bek and Jakobsson 1994), sodium channels (Krishnamurthy et al. 2007) and their interactions with blockers (Gordon and Chung 2012), properties of the KcsA potassium channel (Chung et al. 1999) and its interaction with blockers (Cui et al. 2001; Bisset and Chung 2008), ionic channels from purinergic

P2X family receptors (Turchenkov and Bystrov 2014), and GABAA receptor membrane channel domain (O'Mara et al. 2005).

PNP model does not require vast computational resources for long-time-scale simulations because the whole system is represented by a continuous medium. This includes the channel components, as well as the solvent and ions. The PNP model is defined by two equations. The Nernst-Planck equation allows calculating the flux density:

$$J = -D\left(\nabla n + n \cdot \nabla F \cdot \frac{ze}{kT}\right) \tag{2.2}$$

and the Poisson equation is used for calculating electric potential in the pore:

$$\varepsilon_0 \nabla \cdot \left[\varepsilon(r) \nabla F(r)\right] = -e\sum_j z_j n_j(r) - \rho(r) \tag{2.3}$$

Here, J is the current density, n is the ionic density, F represents the electric potential, ε_0 is the permittivity of free space, ε is the permittivity of the medium, and ρ represents external charges (from the pore walls).

The studies by Edwards et al. (2002), Noskov et al. (2004), Cozmuta et al. (2005), Dyrka et al. (2008, 2013, and 2016), Wang et al. (2014), Chaudhry et al. (2014), and Liu and Eisenberg (2015) have shown the potential of the PNP model in research of several transmembrane proteins. First, the PNP model was used for calculating ionic conduction through wide biological pores, including phenomena of charge selectivity and rectification. Correct results for wide pores, such as alpha-hemolysin, encouraged development of more elaborate PNP algorithms, which enable approximate studies of narrow ionic channels with selectivity filters. The PNP models have been also employed to study effects of single-point mutations in transmembrane proteins on the ion conductions and influence of blocking molecules bound to the pores.

2.3 Applications of Bioinformatical Modeling of Channels

Bioinformatical modeling of ionic channels opens new perspectives in understanding, preventing, and treating diseases which are related to channelopathies or non-physiological disruption of the cell membrane. Compound libraries of potential ligands for ionic channels have been designed in silico. They are used in computational and functional studies of ionic channels, such as changes of the transmembrane potential and ion concertation. These results may be easily compared to the physiological measurement to assess the impact of potential drugs, which is a big problem of studies regarding ionic channels.

Ionic channels are involved in numerous physiological processes and can cause channelopathies, diseases resulting from their wrong performance. Mutations in over 60 different ionic channel genes are related to human diseases (Ashcroft 2006), and more than 1400 publications regarding channelopathies can be found in PubMed. The majority of channelopathies is related to the cardiac and nervous systems. Often the first observed clinical phenotypes of cardiac channelopathies are sudden death or cardiac arrest (Webster and Berul 2013). Several cardiac channelopathies have been recognized, for example, long QT syndrome and catecholaminergic polymorphic ventricular tachycardia. The neurological channelopathies influence the brain, spinal cord, peripheral nerves, or muscles (Spillane et al. 2016), for example, epilepsy, pain syndromes, and congenital myasthenic syndromes. Moreover, different ionic channels are overexpressed in some cancer cells, which gives new opportunities for cancer drug design (Kunzelmann 2005).

Bioinformatical studies may give insight into molecular mechanisms of blocking ionic channels, which is often used in drug development. Various studies (Gordon and Chung 2011; Chen and Kuyucak 2011) using MD simulations show binding free energies of charybdotoxin with different voltage-gated potassium channels. Kuyucak and Norton (2014) presented the results of docking the toxin ShK to potassium channels. Other studies by Gianti et al. (2016) showed computational docking of known Hv1 inhibitors, 2-guanidinobenzimidazole (2GBI), and their analogues to the Hv1 proton channel. The recent study by Elokely et al. (2016) gave a new insight into the activation process of transient receptor potential, caused by ligands in the cation channel subfamily V member 1 (TRPV1).

The use of the bioinformatical modeling tools is not limited to regular transmembrane proteins of unknown structural or functional characteristics. In silico approach can be applied to other molecules with affinity to lipid membranes, including irregular clusters of peptides that could interact and insert into the cell membranes. Such entities are even harder to study with classical experimental methods due to transient nature of such interactions, their potentially lower stability, higher dynamics, and much lower reproducibility of these structures. They also lack any specific single channel conductance, characteristic of a particular species. On the other hand, these interactions are of great interest to researchers. Nonphysiological disruption of the cell membrane can lead to severe consequences in the cell functioning, allowing abnormal ion fluxes across the membrane. For example, nonphysiological membrane channels could destroy a high gradient of calcium concentration across the membrane and uncontrolled ionic influx into the cell. Calcium ions are very important cell messengers used in a signal transduction; therefore, disruption of the physiological flows severely affects cell functioning, including the cell death. Studies show that amyloid diseases are accompanied by deposits of protein aggregates, forming elongated beta fibrils made from peptide oligomers. The mature fibrils are not dangerous to cells, but the amyloid oligomers are highly cytotoxic. The mechanisms underlying this lethal effect have not been discovered, yet. Amyloid aggregates are involved in many diseases, often neurodegenerative, such as Alzheimer's disease, Parkinson's disease, amyotrophic lateral sclerosis (ALS), Huntington's disease, and many others. Some cellular effects of these diseases are very similar in different diseases, and they include calcium dysregulation and depolarization of mitochondria. It was also observed that the amyloid oligomers readily interact with cell membranes, often forming heterogeneous membrane structures with conductive transmembrane pores. Their occurrence is related to the same environmental factors as those promoting amyloid deposits, such as high protein concentration, low pH, proteolysis, and high temperature. The pores are slightly cation selective although no specific selectivity has been observed. Very little is known of their structures, and experimental studies do not provide reproducible and clear answers, most probably due to the heterogeneity of the assembles forming the pores. The toxicity of such structures may resemble the toxicity of channel-forming toxins known from the microbial world. Modeling amyloids and their interactions with the cell membranes can help to address several questions regarding the role of amyloidogenic peptides in the amyloid diseases. It may help to establish the affinity of various amyloidogenic oligomers to the cell membrane and methods to downregulate their insertion and resolve the structures and dynamics of the aggregates interacting with the membranes. Modeling would enable predicting pore effects on ionic flows and cell functioning and help to design therapeutic pore blockers. Several bioinformatical studies attempted to address these questions. The most frequently used modeling method has been MD. The first theoretical models of an amyloid channel were proposed by Durell et al. (1994). The authors provided a model explaining the variable conduction states observed from a single channel and showing how lipid headgroups could affect the ion selectivity. More advanced attempts followed later, as reviewed in Jang et al. (2014). Other MD models showed that also alpha helical structures could be involved in the process of amyloid insertion into the membranes (Pannuzzo et al. 2013). Threading or homology methods, possible if a more regular molecular structure is experimentally solved, could be used to test how mutations affect the amyloid channels. On the other hand, modeling topology and affinity of the aggregates to the membrane can show if lethal amyloid structures are always involved in uncontrolled transmembrane conductivity (Zulpo and Kotulska 2015).

2.4 Summary

Ionic channels are necessary for transmembrane transport, which is crucial for proper functioning of the cells and the whole organism. Difficulties in experimental studies regarding the channels resulted in a very low percent of high-resolution structures available in the databases. Therefore, bioinformatical studies may be a good option to obtain lacking information. Modeling can provide the approximate structure of a protein, its affinity to the membrane, as well as finer details of the selectivity filters, gating mechanisms, and potential ligand binding. It is also very useful in modeling irregular transmembrane pores. Some of this information could be sufficient to discover mechanisms of diseases related to improper ionic transports and indicate potential treatments. Bioinformatical tools are based on a great variety of approaches, including de novo, comparative, or homology-based methods. Algorithms make use of knowledge- or physics-based potentials for energy calculation, using a variety of scoring functions and force fields. Statistical modeling employs a variety of machine-learning methods. Functional modeling enables predicting ionic flows and their intracellular consequences.

Acknowledgments M. Kurczyńska would like to acknowledge the funding from "Diamond Grant" DI2011 002141. The work was also partly funded by the Statutory Funds of Wroclaw University of Science and Technology.

References

Adamian L, Nanda V, DeGrado WF, Liang J (2005) Empirical lipid propensities of amino acid residues in multispan alpha helical membrane proteins. Proteins 59:496–509

Aityan SK, Chizmadzhev Y (1986) Simulation of molecular dynamics of water movement in ion channels. Gen Physiol Biophys 5(3):213–229

Alford RF, Koehler Leman J, Weitzner BD, Duran AM, Tilley DC, Elazar A, Gray JJ (2015) An integrated framework advancing membrane protein modeling and design. PLoS Comput Biol 11(9):e1004398

Ashcroft FM (2006) From molecule to malady. Nature 440:440–447

Bagos PG, Liakopoulos TD, Spyropoulos IC, Hamodrakas SJ (2004) PRED-TMBB: a web server for predicting the topology of β-barrel outer membrane proteins. Nucleic Acids Res 32:W400–W404

Barth P, Schonbrun J, Baker D (2007) Toward high-resolution prediction and design of transmembrane helical protein structures. Proc Natl Acad Sci 104:15682–15687

Bek S, Jakobsson E (1994) Brownian dynamics study of a multiply-occupied cation channel: application to understanding permeation in potassium channels. Biophys J 66(4):1028–1038

Berendsen HJC, van der Spoel D, van Drunen R (1995) GROMACS: a message-passing parallel molecular dynamics implementation. Comput Phys Commun 91(1–3):43–56

Berman HM, Westbrook J, Feng Z, Gilliland G, Bhat TN, Weissig H, Shindyalov IN, Bourne PE (2000) The protein data bank. Nucleic Acids Res 28:235–242

Bernsel A, Viklund H, Falk J, Lindahl E, von Heijne G, Elofsson A (2008) Prediction of membrane-protein topology from first principles. PNAS 105(20):7177–7181

Bernsel A, Viklund H, Hennerdal A, Elofsson A (2009) TOPCONS: consensus prediction of membrane protein topology. Nucleic Acids Res 37:W465–W468

Berti C, Furini S, Gillespie D, Boda D, Eisenberg RS, Sangiorgi E, Fiegna C (2014) Three-dimensional Brownian dynamics simulator for the study of ion permeation through membrane pores. J Chem Theory Comput 10(8):2911–2926

Biasini M, Bienert S, Waterhouse A, Arnold K, Studer G, Schmidt T, Kiefer F, Cassarino TG, Bertoni M, Bordoli L, Schwede T (2014) SWISS-MODEL: modelling protein tertiary and quaternary structure using evolutionary information. Nucleic Acids Res 42(W1):W252–W258

Bigelow H, Rost B (2006) PROFtmb: a web server for predicting bacterial transmembrane beta barrel proteins. Nucleic Acids Res 34:W186–W188

Bisset D, Chung SH (2008) Efficacy of external tetrae-thylammonium block of the KcsA potassium channel: molecular and Brownian dynamics studies. Biochim Biophys Acta 1778(10):2273–2282

Blake CCF, Koenig DF, Mair GA, North ACT, Phillips DC, Sarma VR (1965) Structure of hen egg-white lysozyme. A three-dimensional Fourier synthesis at 2 Å resolution. Nature 206:757–761

Brooks BR, Bruccoleri RE, Olafson BD, States DJ, Swaminathan S, Karplus M (1983) CHARMM: a program for macromolecular energy, minimization, and dynamics calculations. J Comput Chem 4:187–217

Brunger AT (2007) Version 1.2 of the crystallography and NMR system. Nat Protoc 2(11):2728–2733

Burykin A, Kato M, Warshel A (2003) Exploring the origin of the ion selectivity of the KcsA potassium channel. Proteins 52(3):412–426
Callebaut I, Hoffmann B, Lehn P, Mornon JP (2017) Molecular modelling and molecular dynamics of CFTR. Cell Mol Life Sci 74(1):3–22
Chang J-M, Di Tommaso P, Taly J-F, Notredame C (2012) Accurate multiple sequence alignment of transmembrane proteins with PSI-Coffee. BMC Bioinformatics 13(Suppl 4):S
Chaudhry JH, Comer J, Aksimentiev A, Olson LN (2014) A stabilized finite element method for modified Poisson-Nernst-Planck equations to determine ion flow through a nanopore. Commun Comput Phys 15(1):93–125
Chen PC, Kuyucak S (2011) Accurate determination of the binding free energy for KcsA-Charybdotoxin complex from the potential of mean force calculations with restraints. Biophys J 100:2466–2474
Chen KYM, Sun J, Salvo JS, Baker D, Barth P (2014) High-resolution modeling of transmembrane helical protein structures from distant homologues. PLoS Comput Biol 10(5):e1003636
Chiu SW, Pandit SA, Scott HL, Jakobsson E (2009) An improved united atom force field for simulation of mixed lipid bilayers. J Phys Chem B 113:2748–2763
Chung SH, Kuyucak S (2002) Ion channels: recent progress and prospects. Eur Biophys J 31:283–293
Chung SH, Allen TW, Hoyles M, Kuyucak S (1999) Permeation of ions across the potassium channel: Brownian dynamics studies. Biophys J 77(5):2517–2533
Clote P, Waldispuhl J, Berger B, Steyaert JM (2006) transFold: a web server for predicting the structure and residue contacts of transmembrane beta-barrels. Nucleic Acids Res 34:W189–W193
Coalson RD, Cheng MH (2010) Discrete-state representation of ion permeation coupled to fast gating in a model of ClC chloride channels: comparison to multi-ion continuous space Brownian dynamics simulations. J Phys Chem B 114(3):1424–1433
Cornell WD, Cieplak P, Bayly CI, Gould IR, Merz KM, Ferguson DM, Spellmeyer DC, Fox T, Caldwell JW, Kollman PA (1995) A second generation force field for the simulation of proteins, nucleic acids, and organic molecules. J Am Chem Soc 117:5179–5197
Cowen L, Bradley P, Menke M, King J, Berger B (2002) Predicting the beta-helix fold from protein sequence data. J Comput Biol 9:261–276
Cozmuta I, O'Keeffe JT, Bose D, Stolc V (2005) Hybrid MD-Nernst Planck model of α-hemolysin conductance properties. Mol Simul 31:79–93
Cui M, Shen J, Briggs JM, Luo X, Tan X, Jiang H, Chen K, Ji R (2001) Brownian dynamics simulations of interaction between scorpion toxin Lq2 and potassium ion channel. Biophys J 80(4):1659–1669
Delemotte L, Klein ML, Tarek M (2012) Molecular dynamics simulations of voltage-gated cation channels: insights on voltage-sensor domain function and modulation. Front Pharmacol 3:97
Dlugosz M, Zielinski P, Trylska J (2011) Brownian dynamics simulations on CPU and GPU with BD_BOX. J Comput Chem 32:2734–2744
Dobson L, Remenyi I, Tusnady GE (2015) CCTOP: a consensus constrained TOPology prediction web server. Nucleic Acids Res 43:W408–W412
Domanski J, Stansfeld PJ, Sansom MSP, Beckstein O (2010) Lipidbook: a public repository for force-field parameters used in membrane simulations. J Membr Biol 236:255–258
Doyle DA, Morais Cabral J, Pfuetzner RA, Kuo A, Gulbis JM, Cohen SL, Chait BT, MacKinnon R (1998) The structure of the potassium channel: molecular basis of K+ conduction and selectivity. Science 280:69–77
Durell SR, Guy HR, Arispe N, Rojas E, Pollard HB (1994) Theoretical models of the ion channel structure of amyloid beta-protein. Biophys J 67(6):2137–2145
Dyrka W, Augousti AT, Kotulska M (2008) Ion flux through membrane channels-an enhanced algorithm for the Poisson-Nernst-Planck model. J Comput Chem 29:1876–1888
Dyrka W, Bartuzel MM, Kotulska M (2013) Optimization of 3D Poisson-Nernst-Planck model for fast evaluation of diverse protein channels. Proteins 81(10):1802–1822

Dyrka W, Kurczyńska M, Konopka BM, Kotulska M (2016) Fast assessment of structural models of ion channels based on their predicted current-voltage characteristics. Proteins 84(2):217–231
Edwards S, Corry B, Kuyucak S, Chung SH (2002) Continuum electrostatics fails to describe ion permeation in the gramicidin channel. Biophys J 83(3):1348–1360
Elokely K, Velisetty P, Delemotte L, Palovcak E, Klein ML, Rohacs T, Carnevale V (2016) Understanding TRPV1 activation by ligands: insights from the binding modes of capsaicin and resiniferatoxin. PNAS 113(2):E137–E145
Fischer W, Brickmann J (1983) Ion-specific diffusion rates through transmembrane protein channels. A molecular dynamics study. Biophys Chem 18(4):323–337
Galperin MY, Fernández-Suárez XM, Rigden DJ (2017) The 24th annual nucleic acids research database issue: a look back and upcoming changes. Nucleic Acids Res 45:D1–D11
Gianti E, Delemotte L, Klein ML, Carnevale V (2016) On the role of water density fluctuations in the inhibition of a proton channel. PNAS 113(52):E8359–E8368
Gnanasambandam R, Ghatak C, Yasmann A, Nishizawa K, Sachs F, Ladokhin AS, Sukharev SI, Suchyna TM (2017) GsMTx4: mechanism of inhibiting mechanosensitive ion channels. Biophys J 112(1):31–45
Gordon D, Chung SH (2011) Permeation and block of the Kv1.2 channel examined using brownian and molecular dynamics. Biophys J 101:2671–2678
Gordon D, Chung SH (2012) Extension of Brownian dynamics for studying blockers of ion channels. J Phys Chem B 116(49):14285–14294
Hayat S, Elofsson A (2012) BOCTOPUS: improved topology prediction of transmembrane β barrel proteins. Bioinformatics 28(4):516–522
Hayat S, Sander C, Marks DS, Elofsson A (2015) All-atom 3D structure prediction of transmembrane β-barrel proteins from sequences. PNAS 112(17):5413–5418
Hayat S, Peters C, Shu N, Tsirigos KD, Elofsson A (2016) Inclusion of dyad-repeat pattern improves topology prediction of transmembrane β-barrel proteins. Bioinformatics 32(10):1571–1573
Hill JR, Deane CM (2013) MP-T: improving membrane protein alignment for structure prediction. Bioinformatics 29(1):54–61
Hinard V, Britan A, Schaeffer M, Zahn-Zabal M, Thomet U, Rougier JS, Bairoch A, Abriel H, Gaudet P (2017) Annotation of functional impact of voltage-gated sodium channel mutations. Hum Mutat 38:485–493
Hulse RE, Sachleben JR, Wen P-C, Moradi M, Tajkhorshid E, Perozo E (2014) Conformational dynamics at the inner gate of KcsA during activation. Biochemistry 53:2557–2559
Im W, Roux B (2002) Ion permeation and selectivity of OmpF porin: a theoretical study based on molecular dynamics, Brownian dynamics, and continuum electrodiffusion theory. J Mol Biol 322(4):851–869
Jang H, Arce FT, Ramachandran S, Kagan BL, Lal R, Nussinov R (2014) Disordered amyloidogenic peptides may insert into the membrane and assemble into common cyclic structural motifs. Chem Soc Rev 43(19):6750–6764
Jernigan RL, Bahar I (1996) Structure-derived potentials and protein simulations. Curr Opin Struct Biol 6:195–209
Joh NH, Wang T, Bhate MP, Acharya R, Wu Y, Grabe M, Hong M, Grigoryan G, DeGrado WF (2014) De novo design of a transmembrane Zn^{2+}-transporting four-helix bundle. Science 346:1520–1524
Jones DT, Buchan DW, Cozzetto D, Pontil M (2012) PSICOV: precise structural contact prediction using sparse inverse covariance estimation on large multiple sequence alignments. Bioinformatics 28(2):184–190
Jorgensen WL, Maxwell DS, Tirado-Rives J (1996) Development and testing of the OPLS all-atom force field on conformational energetics and properties of organic liquids. J Am Chem Soc 118:11225–11236
Kall L, Krogh A, Sonnhammer ELL (2007) Advantages of combined transmembrane topology and signal peptide prediction – the Phobius web server. Nucleic Acids Res 35:W429–W432

Kappas U, Fischer W, Polymeropoulos EE, Brickmann J (1985) Solvent effects in ionic transport through transmembrane protein channels. J Theor Biol 112(3):459–464
Kelley LA, Mezulis S, Yates CM, Wass MN, Sternberg MJ (2015) The Phyre2 web portal for protein modeling, prediction and analysis. Nat Protoc 10(6):845–858
Kelm S, Shi J, Deane CM (2010) MEDELLER: homology-based coordinate generation for membrane proteins. Bioinformatics 26(22):2833–2840
Kendrew JC, Bodo G, Dintzis HM, Parrish RG, Wyckoff H, Phillips DC (1958) A three-dimensional model of the myoglobin molecule obtained by x-ray analysis. Nature 181:662–666
Klauda JB, Venable RM, Freites JA, O'Connor JW, Tobias DJ, Mondragon-Ramirez C, Vorobyov I, MacKerell AD, Pastor RW (2010) Update of the CHARMM all-atom additive force field for lipids: validation on six lipid types. J Phys Chem B 114:7830–7843
Koehler Leman J, Ulmschneider MB, Gray JJ (2015) Computational modeling of membrane proteins. Proteins 83(1):1–24
Konopka BM, Dyrka W, Nebel JC, Kotulska M (2009) Accuracy in predicting secondary structure of ionic channels. Stud Comp Intell 244:315–326
Konopka BM, Ciombor M, Kurczynska M, Kotulska M (2014) Automated procedure for contact-map-based protein structure reconstruction. J Membr Biol 247(5):409–420
Kortemme T, Morozov AV, Baker D (2003) An orientation-dependent hydrogen bonding potential improves prediction of specificity and structure for proteins and protein-protein complexes. J Mol Biol 326:1239–1259
Kozma D, Tusnady GE (2015) TMFoldRec: a statistical potential-based transmembrane protein fold recognition tool. BMC Bioinformatics 16:201
Kozma D, Simon I, Tusnady GE (2013) PDBTM: protein data bank of transmembrane proteins after 8 years. Nucleic Acids Res 41:D524–D529
Krammer EM, Homble F, Prevost M (2013) Molecular origin of VDAC selectivity towards inorganic ions: a combined molecular and Brownian dynamics study. Biochim Biophys Acta 1828(4):1284–1292
Krishnamurthy V, Vora T, Chung SH (2007) Adaptive Brownian dynamics for shape estimation of sodium ion channels. J Nanosci Nanotechnol 7(7):2273–2282
Krogh A, Larsson B, von Heijne G, Sonnhammer EL (2001) Predicting transmembrane protein topology with a hidden Markov model: application to complete genomes. J Mol Biol 305(3):567–580
Kunzelmann K (2005) Ion channels and cancer. J Membr Biol 205:159–173
Kurczynska M, Kotulska M (2014) Ion Move Brownian Dynamics (IMBD) – simulations of ion transport. Acta Bioeng Biomech 16(4):107–116
Kuyucak S, Norton RS (2014) Computational approaches for designing potent and selective analogs of peptide toxins as novel therapeutics. Future Med Chem 6(15):1645–1658
Landau EM, Rosenbusch JP (1996) Lipidic cubic phases: a novel concept for the crystallization of membrane proteins. PNAS 93(25):14532–14535
Lazaridis T (2003) Effective energy function for proteins in lipid membranes. Proteins 52:176–192
Lazaridis T, Karplus M (1999) Effective energy function for proteins in solution. Proteins 35:133–152
Leaver-Fay A, O'Meara MJ, Tyka M, Jacak R, Song Y, Kellogg EH, Thompson J, Davis IW, Pache RA, Lyskov S, Gray JJ, Kortemme T, Richardson JS, Havranek JJ, Snoeyink J, Baker D, Kuhlman B (2013) Scientific benchmarks for guiding macromolecular energy function improvement. Methods Enzymol 523:109
Lee KI, Jo S, Rui H, Egwolf B, Roux B, Pastor RW, Im W (2012) Web interface for Brownian dynamics simulation of ion transport and its applications to beta-barrel pores. J Comput Chem 33(3):331–339
Liu JL, Eisenberg B (2015) Numerical methods for a Poisson-Nernst-Planck-Fermi model of biological ion channels. Phys Rev E Stat Nonlinear Soft Matter Phys 92(1):012711
Liu Z, Xu Y, Tang P (2005) Molecular dynamics simulations of C2F6 effects on gramicidin a: implications of the mechanisms of general anesthesia. Biophys J 88(6):3784–3791

Lomize MA, Pogozheva ID, Joo H, Mosberg HI, Lomize AL (2012) OPM database and PPM web server: resources for positioning of proteins in membranes. Nucleic Acids Res 40:D370–D376

Lomize AL, Lomize MA, Krolicki SR, Pogozheva ID (2016) Membranome: a database for proteome-wide analysis of single-pass membrane proteins. Nucleic Acids Res 45(D1):D250–D255

Maciejewski A, Pasenkiewicz-Gierula M, Cramariuc O, Vattulainen I, Rog T (2014) Refined OPLS all-atom force field for saturated phosphatidylcholine bilayers at full hydration. J Phys Chem B 118:4571–4581

MacKerell AD, Bashford D, Bellott M, Dunbrack RL, Evanseck JD, Field MJ, Fischer S, Gao J, Guo H, Ha S, Joseph-McCarthy D, Kuchnir L, Kuczera K, Lau FT, Mattos C, Michnick S, Ngo T, Nguyen DT, Prodhom B, Reiher WE, Roux B, Schlenkrich M, Smith JC, Stote R, Straub J, Watanabe M, Wiórkiewicz-Kuczera J, Yin D, Karplus M (1998) All-atom empirical potential for molecular modeling and dynamics studies of proteins. J Phys Chem 102:3586–3616

MacKerell AD, Banavali N, Foloppe N (2000) Development and current status of the CHARMM force field for nucleic acids. Biopolymers 56:257–265

Marks DS, Colwell LJ, Sheridan R, Hopf TA, Pagnani A, Zecchina R, Sander C (2011) Protein 3D structure computed from evolutionary sequence variation. PLoS One 6(12):e28766

Marsico A, Scheubert K, Tuukkanen A, Henschel A, Winter C, Winnenburg R, Schroeder M (2010) MeMotif: a database of linear motifs in alpha-helical transmembrane proteins. Nucleic Acids Res 38:D181–D189

Medovoy D, Perozo E, Roux B (2016) Multi-ion free energy landscapes underscore the microscopic mechanism of ion selectivity in the KcsA channel. Biochim Biophys Acta 1858(7 Pt B):1722–1732

Meng XY, Liu S, Cui M, Zhou R, Logothetis DE (2016) The molecular mechanism of opening the helix bundle crossing (HBC) gate of a Kir channel. Sci Rep 6:29399

Molina ML, Giudici AM, Poveda JA, Fernandez-Ballester G, Mon-toya E, Renart ML, Fernandez AM, Encinar JA, Riquelme G, Morales A, Gonzalez-Ros JM (2015) Competing lipid-protein and protein-protein interactions determine clustering and gating patterns in the potassium channel from streptomyces lividans (KcsA). J Biol Chem 290(42):25745–25755

Monticelli L, Robertson KM, MacCallum JL, Tieleman DP (2004) Computer simulation of the KvAP voltage-gated potassium channel: steered molecular dynamics of the voltage sensor. FEBS Lett 564(3):325–332

Monticelli L, Kandasamy SK, Periole X, Larson RG, Tieleman DP, Marrink SJ (2008) The MARTINI coarse-grained force field: extension to proteins. J Chem Theory Comput 4:819–834

Muirhead H, Perutz M (1963) Structure of hemoglobin. A three-dimensional fourier synthesis of reduced human hemoglobin at 5.5 Å resolution. Nature 199:633–638

Noskov SY, Im W, Roux B (2004) Ion permeation through the alpha-hemolysin channel: theoretical studies based on Brownian dynamics and Poisson-Nernst-Plank electrodiffusion theory. Biophys J 87(4):2299–2309

Notredame C, Higgins DG, Heringa J (2000) T-Coffee: a novel method for fast and accurate multiple sequence alignment. J Mol Biol 302(1):205–217

Nugent T, Jones DT (2009) Transmembrane protein topology prediction using support vector machines. BMC Bioinformatics 10:159

Nugent T, Jones DT (2012) Accurate de novo structure prediction of large transmembrane protein domains using fragment-assembly and correlated mutation analysis. PNAS 109(24):E1540–E1547

O'Mara M, Cromer B, Parker M, Chung SH (2005) Homology model of the GABAA receptor examined using Brownian dynamics. Biophys J 88(5):3286–3299

Oakes V, Furini S, Domene C (2016) Voltage-gated sodium channels: mechanistic insights from atomistic molecular dynamics simulations. Curr Top Membr 78:183–214

Oostenbrink C, Villa A, Mark AE, van Gunsteren WF (2004) A biomolecular force field based on the free enthalpy of hydration and solvation: the GROMOS force-field parameter sets 53A5 and 53A6. J Comput Chem 25:1656–1676

Pannuzzo M, Raudino A, Milardi D, La Rosa C, Karttunen M (2013) α-helical structures drive early stages of self-assembly of amyloidogenic amyloid polypeptide aggregate formation in membranes. Sci Rep 3:2781

Phillips JC, Braun R, Wang W, Gumbart J, Tajkhorshid E, Villa E, Chipot C, Skeel RD, Kale L, Schulten K (2005) Scalable molecular dynamics with NAMD. J Comput Chem 26:1781–1802

Pirovano W, Feenstra KA, Heringa J (2008) PRALINETM: a strategy for improved multiple alignment of transmembrane proteins. Bioinformatics 24(4):492–497

Pluhackova K, Bockmann RA (2015) Biomembranes in atomistic and coarse-grained simulations. J Phys Condens Matter 27:32310

Randall A, Cheng J, Sweredoski M, Baldi P (2008) TMBpro: secondary structure, beta-contact and tertiary structure prediction of transmembrane beta-barrel proteins. Bioinformatics 24:513–520

Restrepo-Angulo I, Vizcaya-Ruizb A, Camacho J (2010) Ion channels in toxicology. J Appl Toxicol 30:497–512

Roux B, Karplus M (1994) Molecular dynamics simulations of the gramicidin channel. Annu Rev Biophys Biomol Struct 23:731–761

Roy Choudhury A, Novic M (2015) PredβTM: a novel β-transmembrane region prediction algorithm. PLoS One 10(12):e0145564

Saier MH, Reddy VS, Tsu BV, Ahmed MS, Li C, Moreno-Hagelsieb G (2016) The transporter classification database (TCDB): recent advances. Nucleic Acids Res 44:D372–D379

Schirmer T, Phale PS (1999) Brownian dynamics simulation of ion flow through porin channels. J Mol Biol 294(5):1159–1167

Seyler S, Beckstein O (2014) Sampling large conformational transitions: adenylate kinase as a testing ground. Mol Simul 40:855–877

Shaw DE (2005) A fast, scalable method for the parallel evaluation of distance-limited pairwise particle interactions. J Comput Chem 26:1318–1328

Shaw DE, Maragakis P, Lindorff-Larsen K, Piana S, Dror RO, Eastwood MP, Bank JA, Jumper JM, Salmon JK, Shan Y, Wriggers W (2010) Atomic-level characterization of the structural dynamics of proteins. Science 330:341–346

Shen H, Chou JJ (2008) MemBrain: improving the accuracy of predicting transmembrane helices. PLoS One 3(6):e2399

Simons KT, Kooperberg C, Huang E, Baker D (1997) Assembly of protein tertiary structures from fragments with similar local sequences using simulated annealing and Bayesian scoring functions. J Mol Biol 268(1):209–225

Siu SWI, Pluhackova K, Bockmann RA (2012) Optimization of the OPLS-AA force field for long hydrocarbons. J Chem Theory Comput 8:1459–1470

Soding J, Biegert A, Lupas AN (2005) The HHpred interactive server for protein homology detection and structure prediction. Nucleic Acids Res 33:W244–W248

Song Y, Tyka M, Leaver-Fay A, Thompson J, Baker D (2011) Structure-guided forcefield optimization. Proteins 79:1898–1909

Spillane J, Kullmann DM, Hanna MG (2016) Genetic neurological channelopathies: molecular genetics and clinical phenotypes. J Neurol Neurosurg Psychiatry 87:37–48

Stansfeld PJ, Goose JE, Caffrey M, Carpenter EP, Parker JL, Newstead S, Sansom MSP (2015) MemProtMD: automated insertion of membrane protein structures into explicit lipid membranes. Structure 23(7):1350–1361

Taylor PD, Toseland CP, Attwood TK, Flower DR (2006) Beta barrel trans-membrane proteins: enhanced prediction using a Bayesian approach. Bioinformation 1(6):231–233

Thomas PD, Dill KA (1996) An iterative method for extracting energy-like quantities from protein structures. PNAS 93:11628–11633

Tieleman D (2012) Computer Simulation of Membrane Dynamics. In: Egelman EH (ed) Comprehensive biophysics. Academic, Oxford, pp 312–336

Tsaousis GN, Tsirigos KD, Andrianou XD, Liakopoulos TD, Bagos PG, Hamodrakas SJ (2010) ExTopoDB: a database of experimentally derived topological models of transmembrane proteins. Bioinformatics 26:2490–2492

Turchenkov DA, Bystrov VS (2014) Conductance simulation of the purinergic P2X2, P2X4, and P2X7 ionic channels using a combined Brownian dynamics and molecular dynamics approach. J Phys Chem B 118(31):9119–9127

Tusnady GE, Simon I (2001) The HMMTOP transmembrane topology prediction server. Bioinformatics 17(9):849–850

Tusnady GE, Kalmar L, Simon I (2008a) TOPDB: topology data bank of transmembrane proteins. Nucleic Acids Res 36:D234–D239

Tusnady GE, Kalmar L, Hegyi H, Tompa P, Simon I (2008b) TOPDOM: database of domains and motifs with conservative location in transmembrane proteins. Bioinformatics 24:1469–1470

Ujwal R, Abramson J (2012) High-throughput crystallization of membrane proteins using the Lipidic Bicelle method. J Vis Exp 59:e3383

Ulmschneider MB, Ulmschneider JP, Schiller N, Wallace BA, von Heijne G, White SH (2014) Spontaneous transmembrane helix insertion thermodynamically mimics translocon-guided insertion. Nat Commun 5:4863

Viklund H, Elofsson A (2004) Best α-helical transmembrane protein topology predictions are achieved using hidden Markov models and evolutionary information. Protein Sci 13(7):1908–1917

Viklund H, Elofsson A (2008) OCTOPUS: improving topology prediction by two-track ANN-based preference scores and an extended topological grammar. Bioinformatics 24(15):1662–1668

Waldispuhl J, O'Donnell CW, Devadas S, Clote P, Berger B (2008) Modeling ensembles of transmembrane beta-barrel proteins. Proteins 71:1097–1112

Wang H, He Z, Zhang C, Zhang L, Xu D (2013) Transmembrane protein alignment and fold recognition based on predicted topology. PLoS One 8(7):e69744

Wang J, Zhang M, Zhai J, Jiang L (2014) Theoretical simulation of the ion current rectification (ICR) in nano-pores based on the Poisson-Nernst-Planck (PNP) model. Phys Chem 16(1):23–32

Wanga S, Ladizhansky V (2014) Recent advances in magic angle spinning solid state NMR of membrane proteins. Prog Nucl Magn Reson Spectrosc 82:1–26

Wassenaar TA, Ingolfsson HI, Priess M, Marrink SJ, Schafer LV (2013) Mixing MARTINI: electrostatic coupling in hybrid atomistic coarse-grained biomolecular simulations. J Phys Chem B 117:3516–3530

Webb B, Sali A (2016) Comparative protein structure modeling using MODELLER. Curr Protoc Bioinformatics 54:5.6.1–5.6.37

Webster G, Berul CI (2013) An update on Channelopathies from mechanisms to management. Circulation 127:126–140

Weiner SJ, Kollman PA, Case DA, Singh UC, Ghio C, Alagona G, Profeta S, Weiner P (1984) A new force field for molecular mechanical simulation of nucleic acids and proteins. J Am Chem Soc 106:765–784

Weiner BE, Woetzel N, Karakaş M, Alexander N, Meiler J (2013) BCL::MP-fold: folding membrane proteins through assembly of transmembrane helices. Struct Lond Engl 21:1107–1117

Woetzel N, Karakas M, Staritzbichler R, Muller R, Weiner BE, Meiler J (2012) BCL::score-knowledge based energy potentials for ranking protein models represented by idealized secondary structure elements. PLoS One 7:e49242

Yarov-Yarovoy V, Schonbrun J, Baker D (2006) Multipass membrane protein structure prediction using Rosetta. Proteins Struct Funct Bioinformatics 62:1010–1025

Zhang T, Kolinski A, Skolnick J (2003) TOUCHSTONE:II a new approach to ab initio protein structure prediction. Biophys J 85:1145–1164

Zulpo M, Kotulska M (2015) Comparative modeling of hypothetical amyloid pores based on cylindrin. J Mol Model 21(6):151

Chapter 3
Cell Membrane Transport Mechanisms: Ion Channels and Electrical Properties of Cell Membranes

Julita Kulbacka, Anna Choromańska, Joanna Rossowska, Joanna Weżgowiec, Jolanta Saczko, and Marie-Pierre Rols

Abstract Cellular life strongly depends on the membrane ability to precisely control exchange of solutes between the internal and external (environmental) compartments. This barrier regulates which types of solutes can enter and leave the cell. Transmembrane transport involves complex mechanisms responsible for passive and active carriage of ions and small- and medium-size molecules. Transport mechanisms existing in the biological membranes highly determine proper cellular functions and contribute to drug transport. The present chapter deals with features and electrical properties of the cell membrane and addresses the questions how the cell membrane accomplishes transport functions and how transmembrane transport can be affected. Since dysfunctions of plasma membrane transporters very often are the cause of human diseases, we also report how specific transport mechanisms can be modulated or inhibited in order to enhance the therapeutic effect.

J. Kulbacka (✉) • A. Choromańska • J. Saczko
Department of Medical Biochemistry, Wroclaw Medical University,
Chalubińskiego 10, 50-367 Wrocław, Poland
e-mail: julita.kulbacka@umed.wroc.pl

J. Rossowska
Institute of Immunology and Experimental Therapy Polish Academy of Sciences,
Rudolfa Weigla 12, 53-114 Wrocław, Poland

J. Weżgowiec
Department of Dental Prosthetics, Division of Dental Materials, Wrocław Medical University,
Krakowska 26, 50-425 Wrocław, Poland

M.-P. Rols
Institut de Pharmacologie et de Biologie Structurale, Université de Toulouse, CNRS, UPS,
205 Route de Narbonne, BP 64182, F-31077 Toulouse, France

J. Kulbacka, S. Satkauskas (eds.), *Transport Across Natural and Modified Biological Membranes and its Implications in Physiology and Therapy*,
Advances in Anatomy, Embryology and Cell Biology 227,
DOI 10.1007/978-3-319-56895-9_3

3.1 Introduction

Cell membranes are more than just passive barriers separating cells from their environment; they are dynamic structures. It is commonly known that cell membranes are selectively permeable and allow some substances to pass through while limiting the passage of others. Thus, cell membranes can actively participate in all cellular exchange processes. The uptake of components required for existence and communication among cells and between cells and their environment occurs through the membrane interface (Ashrafuzzaman and Tuszynski 2013). Moreover, cells can control and affect input and export reactions and finally mediate the further signal response. For all therapeutic protocols requiring drug delivery, an effective transport across the cellular membranes is of primary importance. Although hydrophobic small molecules can enter the cell membrane via simple diffusion, most drugs used in cancer therapies need carrier proteins for their transmembrane transport. Thus, many researchers focus on the development of new methods to enhance drug transport and uptake into targeted cells.

3.1.1 Structure and Function of the Plasma Membrane

The plasma membrane forms the boundary of a living cell. However, it is not only a barrier but also allows for transport between the cell and its environment. In the basic structural model of the cell membrane – the fluid mosaic model – the membrane is considered as a ca. 7 nm thin lipid bilayer with integral proteins inserted in the membrane and peripheral proteins bound to the membrane by protein-protein interactions with integral membrane proteins.

The hydrophobic interior of the bilayer, composed of fatty acid chains, makes the membrane impermeable to water-soluble molecules, while hydrophilic head groups are in contact with surrounding water. Due to their amphipathic character, phospholipids can spontaneously form stable bilayers in an aqueous environment in order to minimize the exposure of hydrophobic fatty acid chains. Membranes are highly dynamic, and both lipids and proteins localized in the plasma membrane can rotate or diffuse laterally and are distributed asymmetrically in the two leaflets. The outer leaflet of the plasma membrane contains two predominant phospholipids: phosphatidylcholine and sphingomyelin, as well as glycolipids. The major components of the inner leaflet are phosphatidylethanolamine, phosphatidylserine, and phosphatidylinositol. Additionally, cholesterol is present in the both leaflets and influences the membrane fluidity. Specific domains rich in cholesterol and sphingolipids are called lipid rafts.

The structure of phospholipid bilayer determines the fundamental function of the plasma membrane, which is a selectively permeable barrier separating the inside and the outside of the cell. Membrane proteins play a crucial role in ensuring selective transport of molecules across the membrane as well as controlling interactions between cells. They can act as ion channels, pumps, receptors, enzymes, or energy transducers. Selective permeability of the cell membrane accounts for the regulation of basic cellular processes through maintaining the osmotic pressure and cellular pH, as well as regulation of drug uptake mechanisms (Cooper 2000; Ray et al. 2016).

3.2 Transport Across the Membranes

3.2.1 *Passive Transport*

All cells should adsorb, communicate, and exchange small- and medium-size molecules. The function of transport across the cell membranes is one of the mechanisms which are needed for normal cell function. With respect to its functional role, the cell membrane is semipermeable and selective. The movement of molecular substances and ions across the biological membrane can occur actively or passively. The passive transport comprises movement of solutes along the concentration gradient (high concentration to low concentrations); thus, it does not require energy. We can distinguish few types of this transport: simple diffusion, facilitated diffusion, and osmosis.

All ions and molecules are suspended in aqueous solution. The movement of this water solution across the membrane is called osmosis. Thus, the concentration of all solutes is determined by osmotic concentration, and we can observe cell swelling when, for example, water diffuses into the cell inside from the extracellular fluid. A very important role in osmosis process is attributed to aquaporins (AQPs), i.e., channels which are not specific for water transport but are also able to transport other molecules such as glycerol, urea, or arsenic (Alleva et al. 2012). It is known that eukaryotic aquaporins can be directly regulated by pH, phosphorylation, and divalent cations. Currently, aquaporins are considered as structures implicated in some diseases, and AQP inhibitors are used in a therapeutic approach. Pharmacological AQP modulation has been studied as a potential therapeutic target for human diseases involving water imbalance such as congestive heart failure, hypertension, and glaucoma (Alleva et al. 2012; Gerbeau et al. 2002; Headfalk et al. 2006; Yukutake et al. 2009).

The next type of a passive transport is simple diffusion which occurs when random motion of ions dissolved in water causes a net movement of these substances from the regions of higher concentration to compartments with lower concentration. The process will continue until both concentrations will be equal. This process may be affected by temperature or the size of molecules (Lodish et al. 2000; Sugano et al. 2010).

A more multifaceted type of transport is the facilitated diffusion, which also concerns passive movement of molecules but via protein carriers, permeases, channel proteins, or transporters. This group comprises ionophores, ion channels, or transport proteins. Protein carriers are integral glycoproteins which transport ions, sugars, or amino acids. They are specific for a certain type of solute and enhance transport by physically binding to specific molecules on one side of the membrane and releasing them on the other side. There are more than 400 membrane transporters which belong to two superfamilies of membrane transporters: ATP-binding cassette (ABC) and solute carriers (SLC) families. Some of them are important in clinical practice because of drug disposition and side effects (Albert et al. 2002; Sugano et al. 2010). These carriers can be functionally classified into influx and efflux transporters according to the direction of movement of substrates (Li et al. 2014). Perland et al. described that SLCs are crucial for maintaining homeostasis within the body as they control molecular trafficking across cellular lipid membranes. Genetic polymorphisms in SLCs appear to be associated with several diseases, such as amyotrophic lateral sclerosis (ALS), Alzheimer's disease, and schizophrenia. Moreover, SLCs can function as drug targets and enable drug absorption into specific organs (Perland and Fredriksson 2016; Lin et al. 2015; Rask-Andersen et al. 2013).

3.2.1.1 Ion Channels in Cell Membrane

The knowledge about the ion channels is presented in greater detail, because they play crucial roles in the growth/proliferation, migration, and invasion of tumor cells. Thus, ion channels represent a promising target for the development of novel and effective cancer therapies (Li and Xiong 2011; Kaczorowski et al. 2008; Becchetti et al. 2013).

Ion channels form hydrophilic, highly selective pores allowing passive transport of inorganic ions through the plasma membrane. Ions of appropriate size and charge – primarily Na^+, K^+, Ca^{2+}, or Cl^- – cross the membrane down their electrochemical gradient at a rate that is approximately 1000 times higher than that reached by other transport proteins. Channels for anions have positively and for cations negatively charged side chains in the pore, i.e., the channel pore is charge specific. The second important feature of ion channels is that they could be gated and do not always allow ions to freely diffuse across the membrane. Apart from the high selectivity and efficiency in ion passing, the channels are characterized and simultaneously differentiated by mechanism of closing ("gating"). Under appropriate stimulation, they change their conformation to allow ion flux. Depending on the type of ion channels, they can open and close in response to (Fig. 3.1) (Alberts et al. 2010):

- Changes of voltage or membrane potential (voltage-gated channels)
- Binding of ligands such as neurotransmitters (transmitter-gated channels), ions (ion-gated channels), and nucleotides (nucleotide-gated channels)

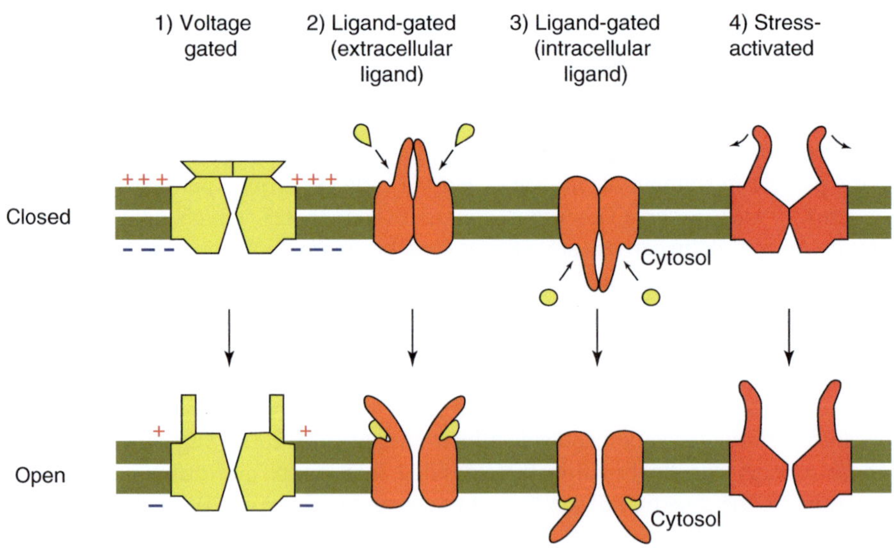

Fig. 3.1 The gating of ion channels. Different incentives that open ion channels (acc. Alberts et al. 2010)

- Mechanical stress – vibration or pressure, such as sound waves or the pressure of touch (mechanically gated channels)
- Light (light-gated channels)

Ion channels are known mainly as mediators of electrical signaling in the nervous system and of electrical excitability of muscle cells. However, they are found in all animal cells, and their ability to control ion fluxes is essential for different cell functions. In this review, we focus our attention on voltage-gated channels due to their sensitivity to depolarization of the plasma membrane.

More than 100 types of ion channels have been described so far and 10 or more kinds of ion channels can be found in the plasma membrane of a single nerve cell.

Voltage-Gated Sodium Channels (Na_V) Voltage-gated sodium channels (Na_V) are described as critical (1) in conducting inward current that depolarizes the plasma membrane and (2) in initiation of action potential in excitable cells. Na_V channels are composed of the single pore-forming α subunit which is associated with two β subunits, uncovalently linked β1 or β3 subunits, and covalently linked β2 or β4 subunits (Brackenbury and Isom 2011). The α subunit is composed of four homologous domains forming the Na^+-conducting pore. Each domain contains six transmembrane segments among which the fourth segment of each domain contains the voltage-sensing domain of the channel. There are nine different α subunits (Na_V1.1–Na_V1.9) which have unique tissue-specific expression patterns. The characteristic future of Na_V is their ability to self-inactivation that occurs despite ongoing depolarization. The α subunit contains the inactivation loop which efficiently blocks ion conduction. On the other hand, β subunits are involved in the modulation of the biophysical properties of the α subunit and act simultaneously as cell adhesion molecule (reviewed by Kruger and Isom 2016). Due to their role in generation of action potentials, Na_V channels are highly expressed in excitable cells including neurons, cardiac myocytes, and skeletal muscle. However, apart from excitable cells, they are also expressed in normal cells including fibroblast, immune cells, progenitor cells, and colon, cervix, or glial cells, where they regulate processes such as proliferation, differentiation, or migration (Brackenbury et al. 2008). Additionally, it has been shown that mutations in Na_V genes or changes in their expression are associated with different diseases. Mutation in Na_V genes is linked with epilepsy, cardiac arrhythmia, neuropathic pain, or migraine (Adsit et al. 2013; O'Malley and Isom 2015), while abnormal expression of Na_V seems to be related to higher motility and metastatic potential of cancer cells and tumor aggressiveness. Patel and Brackenbury proposed the following model for Na_V channel involvement in cancer progression. Primary tumor cells express β subunits contributing to adhesion and promoting angiogenesis and resistance to apoptosis. Additionally, upregulation of α subunits promotes a mesenchymal-like phenotype, activation of proteases, and local invasion from the primary tumor (Patel and Brackenbury 2015).

Voltage-Gated Potassium Channels (K_Vs) Voltage-gated potassium channels (K_Vs) are widely distributed in a variety of cells and are responsible for retrieving and maintaining the original negative potential of the plasma membrane. They open upon depolarization and cause rapid efflux of K^+ to drive membrane back toward K^+ equilibrium potential. Moreover, K_Vs play crucial roles in regulation of cell volume, differentiation, proliferation, migration, and apoptosis (Yellen 2002; Johnston et al. 2010; O'Grady and Lee 2005; Wang et al. 2002). Considering their biophysical properties, K_Vs are divided into twelve subfamilies (K_V1–K_V12). Each K_V channel is a tetramer composed of four α subunits forming the K^+-conduction pore. Like in Na_V channels, each α subunit in K_Vs contains six transmembrane segments with voltage sensor domain and one pore loop comprising five to six segments which form the pore domain. The complete K_V channel often contains accessory β subunits. Due to the possibility to form homotetramers and heterotetramers, there is a high diversity in K_V structure and properties. Members of K_V1–K_V4 subfamilies are responsible for forming functional channels, while K_V5, K_V6, K_V8, and K_V9 are electrically silent and play a modulatory role. They form heterotetramers mainly with K_V2 subfamily modulating their properties (Gutman et al. 2005; McKeown et al. 2008). K_V1, K_V4, and K_V7 belong to low-voltage-activated channels and are responsible for limiting the number of action potentials generated in response to depolarization. Loss of proper function of K_V1 or K_V7 is connected with hyperexcitability phenotypes occurring in episodic ataxia type 1 or epilepsy (Brown and Passmore 2009; Brew et al. 2003). On the other hand, high-voltage-activated K_V2 channels influence action potential duration. Increased efflux of ions through $K_V2.1$ channels promotes apoptotic signaling (Misonou et al. 2005; Mohapatra et al. 2009). Despite their great significance in neuronal excitability, it has been also shown that K_Vs are involved in tumor progression and malignancy. K_V11 and K_V10 channels were among the first voltage-gated channels which were directly related to cancer. Inhibition of these channels in cancer cells contributes to the decrease of proliferation rate and migration and initiates apoptosis (Pardo and Stühmer 2014).

Voltage-Gated Calcium Channels (Ca_Vs) Voltage-gated calcium channels (Ca_Vs) are key mediators of calcium influx into the cell and take part in cell signaling by induction of action potentials in excitable cells as well as by activation of calcium-dependent enzymes. Increasing the intracellular concentration of Ca^{2+} ions, Ca_Vs act as inducers of numerous physiological processes including apoptosis, hormone release, activation of gene transcription, or muscle contraction (Zamponi 2016). Traditionally, Ca_Vs are divided into two main groups: low-voltage-activated (also known as T-type) and high-voltage-activated channels including L-, P-, Q-, and R-subfamilies, where P- and Q-type channels are distinguished by alternative splicing and channel subunit composition. All these subtypes correspond to ten different $Ca_V\alpha1$ subunits which are encoded in the mammalian genome. T-type channels include $Ca_V3.1$, $Ca_V3.2$, and $Ca_V3.3$ α1 subunits, L-type channels include Ca_V1-4 family, while P-, Q-, and R-type channels comprise Ca_V2 family (isoforms 1–3). The main component of the Cav is the α1 subunit composed of four transmembrane domains. Each of the domain contains six transmembrane segments with voltage sense motif allowing the channel to open in response to depolarization and p-loop motif responsible for forming the transmembrane pore and for selectivity for Ca^{2+} ions. The influx of calcium through α1 subunit is modulated by its interaction with accessory subunits $Ca_V\beta$ (4 isoforms $Ca_V\beta1$-4), $Ca_V\alpha2$-δ (4 isoforms $Ca_V\alpha2$-δ1-4), and $Ca_V\gamma$ (8 isoforms $Ca_V\gamma1$-8). The interaction of these subunits with α1 significantly accelerates the activation and deactivation kinetics and increases the maximal conductance of ionic current (Simms and Zamponi 2014; Catterall et al. 2005). Taking into consideration the diversity of particular subunits and their capability to interact with each other, a vast number of Ca_V channels can be generated which show different properties and functionality. Ca_V dysregulation in excitable cells is connected with numerous disorders such as pain, Parkinson disease, epilepsy, or hypertension. It was shown that expression of T-type Ca_V channels decreases during cell development and differentiation indicating the involvement of Ca_V in proliferation process. Moreover, their increasing activity in normal cells can cause an overload with cytoplasmic calcium triggering apoptosis (Wang et al. 1999). However, it was also observed that inhibition of the ion channels in some cancer cell lines contributes to induction of p53-dependent apoptosis through activation of the p38-MAPK signal pathway (Dziegielewska et al. 2014) or caspase-dependent apoptosis (Das et al. 2013). For these reason T-type Ca_V channels are proposed as an attractive target for antitumor strategies.

Certainly, all voltage-gated channels (VGCs) play important roles in different physiological processes and all alterations in their functioning and properties contribute to disorders both on the level of excitable or normal cells. Activators of VGC channels may be used in human diseases associated with too high membrane excitability including epilepsy or arrhythmia. Several novel activators of K_V7 channel, responsible for controlling membrane excitability, have been validated in clinical trials as potential therapeutic agents for treatment for epilepsy, different types of pain, or neurodegenerative and psychiatric disorders (Xiong et al. 2008). On the other hand, L-type Ca_V channel activators have been reported as inducers of cardiac arrhythmias and elevated arterial blood pressure. They initiated also increased release of neurotransmitters as well as massive neuronal activation in most brain regions. For these reasons, VGC activators are usually used to study their role in cellular signaling (Zamponi et al. 2015). However, there are many clinically approved drugs blocking VGC activity. Currently used antiepileptic retigabine, antiarrhythmic verapamil, or antidepressant imipramine drugs possess inhibitory activity against potassium channels including $K_V7.1$, $K_V11.1$, or $K_V10.1$, respectively, whereas, mibefradil, nifedipine, pimozide, or derivatives of dihydropyridine applied for treatment of hypertension or schizophrenia and psychosis are blockers of Ca_V channels. Successively, ranolazine or riluzole applied in angina pectoris or amyotrophic lateral sclerosis are known inhibitors of $Na_V1.5$ and $Na_V1.7$ channels (reviewed by Kale et al. 2015). It has been shown that almost all abovementioned drugs also possess anticancer activity. Imipramine is effective in inhibition of melanoma cells (Gavrilova-Ruch et al. 2002) and in induction of apoptosis in ovarian cancer cells (Asher et al. 2011). T-type calcium channel blockers arrest the cell cycle in G1 or S phase sensitizing cancer cells to cytostatic drugs (Rim et al. 2012). Blocking L-type Ca_V channels by nifedipine is connected with reduction of mitogenic effect of endothelin 1 in lung cancer (Zhang et al. 2008). Verapamil, in turn, has elicited antiproliferative effect in murine breast cancer (Taylor and Simpson 1992). Moreover, its co-application with ifosfamide prolonged survival of patients suffering from non-small lung cancer (Millward et al. 1993). Apart from chemical compounds, a prolonged inhibition of voltage-gated (VG) currents through Na^+, Ca^{2+}, and K^+ channels was observed after permeabilization of the cell plasma membrane by intense nsPEFs (Pakhomov et al. 2007a, b; Bowman et al. 2008; Nesin et al. 2012). Thus, nanosecond duration nanopulses seem to be effective inhibitors of voltage-gated sodium or calcium channels. Although different ways for blocking VGCs activity were discovered, it is still necessary to develop new generation of VGC inhibitors with higher selectivity to minimize side effects associated with their application.

3.2.2 Active Transport

The active type of transport (AT) enables movement of solutes against a concentration gradient (from low concentration to high concentration). This is possible because of the used expenditure of energy (ATP hydrolysis). There are two main types of active transport: primary (direct) active transport which directly uses ATP for energy and secondary (indirect) active transport, which is also called cotransport and involves transfer of two distinct molecules (symport, in the same direction; antiport, in opposite directions).

Active transport involves highly selective protein carriers within the membrane and is one of the most significant features of any cell. It allows to take up additional molecules in concentrations higher than in the extracellular fluid. This type of transport is efficiently used in the liver to accumulate glucose molecules from the blood plasma, because the glucose concentration is regularly higher inside the liver cells than it is in the plasma. Moreover, active transport also enables to move substances from the cytoplasm to the extracellular fluid despite higher external concentrations. AT has the vector nature – unidirectional, what means that a metabolite is moved across the membrane in only one direction. AT occurs by specific conveyors – membrane lipoproteins, which have generally two active sites. One of them couples with ATP as energy source. These protein carriers show characteristic ATPase activity (an enzyme that breaks down ATP to ADP and inorganic phosphate). The Ca^{2+}-ATPase and Na^+/K^+-ATPase pumps are important examples that mediate primary active transport.

Ca^{2+}- ATPase Ca^{2+}- ATPase is responsible for Ca^{2+} entry from the extracellular environment into cellular storage compartments and for Ca^{2+} extrusion. This pump is characterized by high-affinity and low-capacity Ca^{2+} (the plasma membrane Ca^{2+} pump (PMCA)). The PMCA pump belongs to the family of P-type ATPases, which are characterized by the temporary conservation of ATP energy (Brini and Carafoli 2011). There are four PMCA isoforms. PMCA1 and PMCA4 are expressed in most tissues, while PMCA2 and PMCA3 are found mainly in the brain, striated muscle, and mammary gland (Brini et al. 2013). PMCAs have no specific peptide or small molecule extracellular ligand, and thus there is no targetable binding site for agonist or antagonist development from the outside of the cell. Most parts of the PMCAs

are located intracellular with only short extracellular loops connecting adjacent transmembrane domains. Therefore, there is a limited access to the protein from the outside for possible drug targeting. There are numerous PMCA pump dysfunctions which are genetically and nongenetically related. The loss, mutation, or inappropriate expression of different PMCAs is associated with pathologies such as hypertension, low bone density, male infertility, hearing loss, and cerebellar ataxia. Hence, a selective modulation of PMCA isoforms has a therapeutic value in planning treatments for different and complex diseases (Strehler 2013). The most frequently used methods to modulate and block PMCA function involve application of La^{3+}, orthovanadate, fluorescein analogs (e.g., carboxyeosin), and substances antagonizing calmodulin action (trifluoperazine, calmidazolium) (Strehler 2013; Cui et al. 2017). Muscella et al. (2011) indicated that a PMCA-selective inhibitor [Pt(O,O′-acac)(γ-acac)(DMS)] rapidly induced apoptosis in MCF-7 cells.

Na^+/K^+- ATPase Na^+/K^+- ATPase maintains the plasma membrane Na^+ and K^+ gradients, which controls cell volume and water distribution on both sides of the membrane. K^+/Na^+ gradients provide the transmission of nervous stimuli and are the driving force of the active transport of sugars and amino acids. Na^+ and K^+ transport and the maintenance of Na^+ and K^+ gradients are energy-consuming processes utilizing 1/3 of the whole ATP pool in the organism (Shattock et al. 2015). There are some cardiotonic steroids (g-strophanthin, digitoxigenin) which inhibit the activity of the Na^+/K^+ pump by selective blocking phosphorylation of the alpha subunit. This leads to Na^+ accumulation inside the cell. High cytosolic concentrations of Na^+ constrain the excretion Ca^{2+} out of the cell by inhibition of Na^+/Ca^{2+} exchanger (NCX). NCX is one of the essential regulators of Ca^{2+} homeostasis in cardiomyocytes and therefore an important modulator of the cardiac contractile function. This mechanism is efficiently used to increase the strength of the heart muscle contraction (Shigekawa and Iwamoto 2001).

H^+/K^+- ATPase Another mechanism efficiently used in active transport is provided by H^+/K^+- ATPase, also known as gastric hydrogen potassium ATPase. H^+/K^+-ATPase couples outward transport of sodium or protons to the inward transport of potassium and is responsible for gastric acid secretion. This ATPase is the main therapeutic target in treatment of acid-related diseases, and there are several known luminal inhibitors enabling the analysis of the luminal vestibule. One group contains the acid-activated covalent, thiophilic proton pump inhibitors, the most effective of the current acid-suppressive drugs. Their binding sites and trypsinolysis allowed identification of all ten transmembrane segments of this ATPase (Shin et al. 2011).

ABC Transporters ATP-binding cassette (ABC) transporters are ubiquitous integral membrane proteins that actively transport molecules across biological membranes, which is critical for most aspects of cell physiology (Dean et al. 2001; Vasiliou et al. 2009; Linton 2007). Until today, 52 human ABC genes approved by HUGO (Human Genome Organization) (Waterbeemd et al. 2006) were identified. Seven subfamilies are classified as ABC transporters (ABCA through ABCG) that are expressed in both normal and malignant cells. They mediate the transport of many substances, e.g., toxins, from the liver, kidneys, and gastrointestinal tract. Moreover they limit permeation of toxins to vital structures, such as the brain, placenta, and testis. ABC transporters contribute to drug resistance via ATP-dependent drug efflux pumps. P-glycoprotein (P-gp), encoded by the MDR1 gene, is an ABC transporter normally involved in the excretion of toxins from cells. It is also responsible for the resistance to many chemotherapeutic agents (Leonard et al. 2003) because several drugs are P-gp substrates. This group comprises adriamycin, daunorubicin, epirubicin-paclitaxel, docetaxel, vinblastine, VP-16, mitoxantrone, or actinomycin D. Several inhibitors of drug efflux which cause a reversal of drug resistance have been identified: verapamil, quinidine, quinine, cyclosporine A, PSC 833, VX-710, LY335979, R101933, OC144-093, and XR9576 (Leonard et al. 2003; Dean et al. 2001). Many data indicate that also natural compounds like catechins, coumarins, or flavonoids, baicalein, icaritin, or biochanin A (Zhang and Morris 2003), can effectively inhibit or modulate P-gp expression (Abdallah et al. 2015). In conclusion, effective blockade or modulation of ABC transporters might play a significant role in affecting the cellular pharmacokinetics of anticancer drugs and in overcoming tumor resistance.

Other Active Transport Processes Other active transport processes include endocytosis and exocytosis. These processes are related to macromolecule transport. Endocytosis is a process by which cells take up macromolecules, particulate substances, and in some cases even other cells. In this process the plasma membrane extends outward and envelops food particles. Three major types of endocytosis are used by cells: phagocytosis, pinocytosis, and receptor-mediated endocytosis. Uptake of an organism or any organic matter is called phagocytosis ("cellular eating"). If the taken-up material is liquid, the process is called pinocytosis ("cellular drinking"), which is a very common process in animal cells (e.g., mammalian egg cells). Carrier-mediated endocytosis is related to transport through specific receptors. Hence, only molecules that can bind to specific receptors on the membrane surface can be transported via this way. A particular receptor detects the presence of a specific molecule and initiates endocytosis, which is very fast. The most popular receptors are LDL molecules transporting cholesterol into the cell, where it can be incorporated into membranes (Yang and Hinner 2015; Alberts et al. 2002; Cleal et al. 2013). The endocytic machinery participates in the generation, propagation, reception, and interpretation of intercellular signals in the context of animal development (Bökel and Brand 2014).

Exocytosis is a reverse process of endocytosis, a discharge of material from vesicles at the cell surface. In animal cells, exocytosis is used for secreting many hormones, neurotransmitters, digestive enzymes, and other substances (Alberts et al. 2002; Yang and Hinner 2015).

Understanding of the macromolecular transport mechanisms such as endocytosis and exocytosis may help to clarify the mechanism of nanoparticle uptake which is essential for safe and efficient therapeutic application. In particular, exocytosis is of high importance because the removal of nanoparticles with cytotoxic drugs from the body is crucial for their biocompatibility. But, endocytosis is also of great importance for the delivering of nanoparticles in the targeted cells (Oh and Park 2014).

3.3 Electrical Properties of Cell Membranes

Biological membranes are electrical insulators due to their phospholipid bilayer structure and are impermeable to ions, unless specific ion channels are temporarily open. Only substances that can pass through the hydrophobic core can diffuse through cell membrane unaided. Charged particles cannot pass through membrane without assistance. Specific transmembrane and channel proteins make this possible. They are necessary to generate a transmembrane potential and an action potential (Alberts et al. 2010; Tombola et al. 2006). Particularly interesting is the Na^+/K^+ pump which maintains concentration gradients. The concentration of Na^+ is higher outside the cell and the concentration of K^+ is higher inside the cell. Consequently, most living cells make use of ions and charged particles to build up a charge across the membrane (Alberts et al. 2010; Gouaux and MacKinnon 2004).

A membrane potential is the result of difference in the electrical charge on the two sides of membrane, caused by active electrogenic pumping (Na^+- K^+ pump) or from passive ion diffusion. The Na^+-K^+ pump has an important role in maintaining an osmotic balance across the cell membrane by keeping the intracellular concentration of Na^+ low. Because of a low concentration of Na^+ inside the cell, other cations have to balance the charge carried by the negatively charged molecules inside the cell (Gouaux and MacKinnon 2004). The main balancing role is played by K^+ ions, which are actively pumped into the cell and move freely or through the K^+ leak channels. The membrane potential is the voltage difference across a membrane due to a slight excess of positive ions on one side and of negative ions on the other. The membrane potential in equilibrium conditions in which there is no net flow of ions across the plasma membrane is called resting membrane potential (Alberts et al. 2010; Gomułkiewicz et al. 2001; Gillespie and Eisenberg 2002; Eisenberg 2013). Quantitative expression that relates the equilibrium ratio of concentrations of an ion on both sides of a permeable membrane to the voltage difference across the membrane is called *Nernst equation:*

$$V = \frac{RT}{zF} \ln \frac{C_o}{C_i}$$

where V = the equilibrium potential in volts (internal minus external potential), C_o and C_i = outside and inside concentrations of the ion, R = the gass constant, T = the absolute temperature, F = Faraday's constant, z = the valence of the ion, and ln = logarithm to the base e (Alberts et al. 2010).

The membrane potential will stay at the resting voltage until something changes. This change may be caused by binding a neurotransmitter to ligand-gated Na^+ channel.

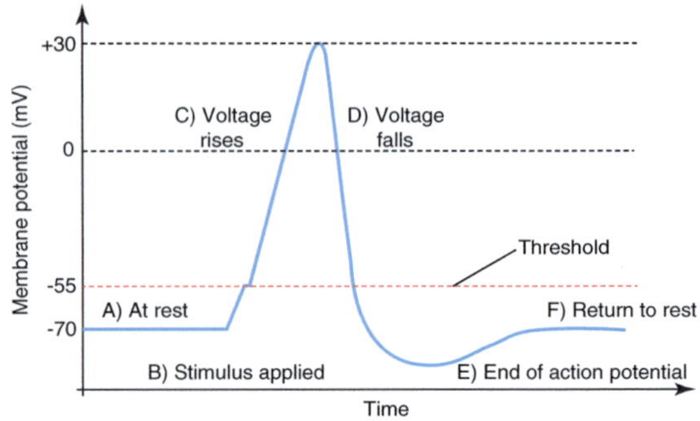

Fig. 3.2 Stages of an action potential (Alberts et al. 2010)

Mechanically gated Na^+ channel will open when a physical stimulus affects a sensory receptor. Sodium starts to enter the cell and the membrane becomes less negative. Another reason for the onset of membrane depolarization is the stimulation of voltage-gated Na^+ channels. Any stimulus that depolarizes the membrane to −55 mV or beyond will cause a large number of channels to open, and an action potential will be initiated. The action potential continues and increases to +30 mV, at which K^+ causes repolarization, including the hyperpolarizing overshoot. As the membrane potential repolarizes and the voltage passes −50 mV again, the channel closes. Potassium continues to leave the cell and the membrane potential becomes more negative, resulting in the hyperpolarizing overshoot. Thereafter the channel closes again and the resting membrane potential is reached because of the activity the Na^+-K^+ pump and the non-gated channels (Alberts et al. 2010; Allen et al. 2006). Figure 3.2 shows stages of an action potential:

All stages of the action potential last approximately 2 ms. While an action potential is in progress, another one cannot be initiated because of the refractory period. The refractory period has two phases: the absolute and the relative refractory period. During first phase, another action potential will not start, because the voltage-gated Na^+ channels are inactivated. When the channels are back to their resting conformation (less than −55 mV), a new action potential could be started but only by a stimulus that is stronger than the former one because K^+ ions flow out of the cell and any Na^+ ions that try to enter will not depolarize the cell but will only prevent hyperpolarization of the cell (Alberts et al. 2010; Gillespie and Eisenberg 2002; Eisenberg 2013).

3.3.1 Plasma Membrane Capacitance

The plasma membrane is an insulating barrier between two conducting media with different concentrations of ions: the cytoplasm and the extracellular space. For this reason it is considered as a capacitor, which maintains a potential difference between two surfaces. To be precise, it is a leaky capacitor, since the ions may be transported across the membrane. The following equation defines the specific capacitance of a membrane, C_m:

$$C_{\mathrm{m}} = \frac{\kappa \varepsilon_0}{d} \left[\frac{\mu F}{\mathrm{cm}^2} \right]$$

where κ is the dielectric constant for the membrane's inner core, ε_0 is the permittivity of free space, and d is the membrane thickness. The value of C_{m} may vary for different types of cells, but typically it is ca. 1.0 μF/cm^2. The total membrane capacitance is proportional to the cell membrane surface area. Therefore, experimental measurements of the membrane capacitance could be used to calculate the cell membrane surface area and to control its changes. The area of the cell membrane surface is also an important determining factor of the activity of neurons (Ashrafuzzaman and Tuszynski 2013; Golowasch and Nadim 2014).

Electrical properties of the cell membranes can be used to monitor cell condition. Various techniques, including patch clamping, electro-rotation, and impedance spectroscopy, have been proposed to measure the capacitance of the cell membranes. The results of different studies focused on the use of the cell membrane capacitance measurements for tumor cell classification are summarized in Table 3.1. Qiao et al. measured impedance of the cell suspensions and extracted the electrical properties of single cells. They revealed that results of the capacitance measurements can be used to distinguish between normal and cancerous breast cells of different stages of carcinogenesis (Qiao et al. 2010). In order to perform impedance spectroscopy measurements of single cells, Tan et al. developed a microfluidic device. They showed that membrane capacitance of two acute myeloid leukemia cell lines (AML2 with a lower metastatic potential and NB4 with a higher metastatic potential) differs significantly (Tan et al. 2012). A similar approach was taken by Zhao et al. on human large-cell lung carcinoma cell lines. They confirmed the utility of cell membrane capacitance measurements for tumor cell classification (Zhao et al. 2014).

Table 3.1 Summary of the results of different studies on the cell membrane capacitance used for tumor cell classification

Cell line	Type of cells	Membrane capacitance [μF/cm^2]	Reference
MCF-10A	Normal breast cells	3.94	Qiao et al. (2010)
MCF-7	Early-stage breast cancer	1.95	
MDA-MB-231	Invasive breast cancer	1.81	
MDA-MB-435S	Late-stage breast cancer	1.10	
AML2	Acute myeloid leukemia	1.69	Tan et al. (2012)
NB4	Acute promyelocytic leukemia	2.25	
95D	High-metastatic human large-cell lung carcinoma	2.00	Zhao et al. (2014)
95C	Low-metastatic human large-cell lung carcinoma	1.62	

3.4 Transport Support

Cell membrane is a crucial factor that determines cell life and efficiency of all drug-dependent therapies. On the basis of the structure, features, and electrical properties of the cell membrane, we can effectively monitor cell condition and molecular transport efficiency. Transport across the cellular membranes is physiologically required, but in some cases, there is a need to enhance this process. The effective support of transport processes can significantly improve therapeutic protocols, where drug delivery and biodistribution is limited. The best effect induced in the cell to increase molecules influx is membrane permeabilization. There are some factors and methods that can increase cell membrane permeability, which are given in the Fig. 3.3.

Electroporation As previously described, the cell membrane has electrical properties. Electrically, the cell plasma membrane can be viewed as a thin protecting sheet surrounded on both sides by aqueous electrolyte solutions. When exposed to a sufficiently strong electric field, the cell membrane will undergo electrical breakdown, which makes it permeable to molecules that are otherwise unable to cross it. The process of rendering the membrane permeable is called membrane electroporation (EP) (Mir et al. 1991; Kotnik et al. 2012). EP method is efficiently used for effective drug (Gehl 2003; Wezgowiec et al. 2013), DNA (Rosazza et al. 2013), or nanoparticles (Kulbacka et al. 2016a, b) delivery.

Sonoporation Ultrasound (US)-induced permeabilization has been employed to enhance the delivery and activity of variety of drugs (Nejad et al. 2016). More details concerning sonoporation were described in Chap. 5 of the book. However, sonoporation involves the treatment of cells in vitro or tissue in vivo with ultrasound in the presence of microbubbles. The exposition of microbubbles to US causes periodic oscillations and collapse, under selected insonation settings. These oscillations

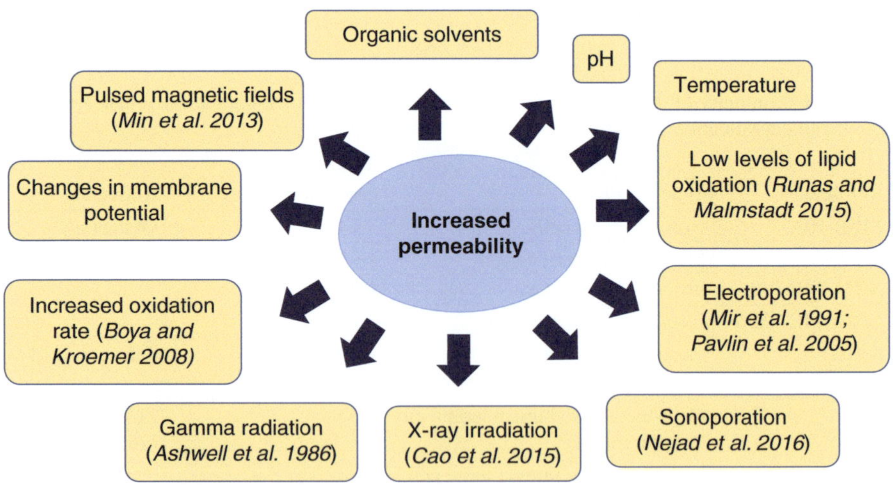

Fig. 3.3 How to increase cell membrane permeability

can induce micro-streaming, shock waves, and/or micro-jets that can affect the integrity of biological membranes (Zeghimi et al. 2015). Sonoporation has been shown to enhance the transport of chemotherapeutics, proteins, genes, and other therapeutic agents into the cell which cannot enter the cell otherwise (Pitt et al. 2004; Fan et al. 2010; Zeghimi et al. 2015). Zeghimi et al. (2015) observed that microbubble destruction probably causes stress of the plasma membrane, with further stimulation of the endocytosis pathways. Thus, the authors have hypothesized that the formation of membrane pores and the stimulation of endocytosis pathways are the main mechanisms in the sonoporation process. Tamošiunas et al. (2016) verified that the combination of electroporation (EP) and ultrasound (US) waves (sonoporation) can result in an increased intracellular delivery of anticancer drugs.

3.5 Summary

The present review describes the basic transport mechanisms across cellular membranes. Transport through membranes is crucial for proper cell function. Each small defect (external factor, mutations, etc.) causes disturbances of this mechanism and finally may cause a number of diseases. Mutations can modify, for example, the physical properties of VGCs and may result in increased sensitivity to voltage changes, which influences conduction and can finally induce certain inherited channelopathies, e.g., epilepsy, long QT syndrome, and paralyses (Delemotte et al. 2010). Transport through cell membranes has attracted the attention of many researchers and stimulated to the development of sensitive and selective ion channel biosensors for high-throughput screening than may be effectively used in modern medical care, drug screening, environmental monitoring, food safety, and biowarfare control (Steller et al. 2012). Sensors of a new generation can be developed on the basis of the membrane differences between cancer and normal cells. The membrane of cancer cells has increased solute transfer of essential nutrients, which is directly involved in the increased proliferation. The application of specific inhibitors or modulators of protein transporters is currently practiced in medicine (Shukla et al. 2011; Selwan et al. 2016). In order to simplify the understanding of the mechanism of transport across the cell membranes, computer modeling is used very often. Several studies focus on the design of new membranes or the improvement of existing ones for simulation of the transport mechanisms (Burganos 2017; Li and Lin 2011; Ateshian et al. 2010).

The available knowledge on the way how metabolites, drug molecules, and ions can be transported enables the control and regulation of the transfer across the membranes and electrical properties of this natural barrier by external factors. The review also describes well-known and new compounds that can modulate or inhibit membrane transporters. Advances in the knowledge of cell membrane transport have been of key importance for an improvement of existing therapies against a wide range of diseases and the design of new approaches.

Acknowledgments This work was supported by the Polish National Science Centre project SONATA BIS 6 - SONB.A040.17.001 (2016/22/E/NZ5/00671).

References

Abdallah HM, Al-Abd AM, El-Dine RS, El-Halawany AM (2015) P-glycoprotein inhibitors of natural origin as potential tumor chemo-sensitizers: a review. J Adv Res 6(1):45–62

Adsit GS, Vaidyanathan R, Galler CM, Kyle JW, Makielski JC (2013) Channelopathies from mutations in the cardiac sodium channel protein complex. J Mol Cell Cardiol 61:34–43

Alberts B, Johnson A, Lewis J, Raff M, Roberts K, Walter P (2002) Molecular biology of the cell, 4th edn. Garland Science, New York. Carrier proteins and active membrane transport. ISBN-10: 0-8153-3218-1

Alberts B, Bray D, Hopkin K, Johnson A, Lewis J, Raff M, Roberts K, Walter P (2010) Essential cell biology. Garland Science, New York

Allen TW, Andersen OS, Roux B (2006) Molecular dynamics potential of mean force calculations as a tool for understanding ion permeation and selectivity in narrow channels. Biophys Chem 124:251–267

Alleva K, Chara O, Amodeo G (2012) Aquaporins: another piece in the osmotic puzzle. FEBS Lett 586(19):2991–2999. ISSN 0014-5793

Asher V, Warren A, Shaw R, Sowter H, Bali A, Khan R (2011) The role of Eag and HERG channels in cell proliferation and apoptotic cell death in SK-OV-3 ovarian cancer cell line. Cancer Cell Int 11:6

Ashrafuzzaman M, Tuszynski JA (2013) Membrane biophysics. Springer-Verlag Berlin Heidelberg, New York, pp 9–30

Ashwell JD, Schwartz RH, Mitchell JB, Russo A (1986) Effect of gamma radiation on resting B lymphocytes. I. Oxygen-dependent damage to the plasma membrane results in increased permeability and cell enlargement. J Immunol 136(10):3649–3656

Ateshian GA, Morrison B, Hung CT (2010) Modeling of active transmembrane transport in a mixture theory framework. Ann Biomed Eng 38(5):1801–1814

Becchetti A, Munaron L, Arcangeli A (2013) The role of ion channels and transporters in cell proliferation and cancer. Frontiers in Physiology 4:312

Bökel C, Brand M (2014) Endocytosis and signaling during development. Cold Spring Harb Perspect Biol 6(3):a017020

Bowman A, Xiao S, Schoenbach KS, Pakhomov AG (2008) Inhibition of voltage-gated calcium channels of cell plasma membrane by nanosecond electric pulses. In: Bioelectromagnetics Society's 30th annual meeting, San Diego, pp 75–76, June 8–12

Boya P, Kroemer G (2008) Lysosomal membrane permeabilization in cell death. Oncogene 27:6434–6451

Brackenbury WJ, Isom LL (2011) Na+ channel beta subunits: overachievers of the ion channel family. Front Pharmacol 2:53

Brackenbury WJ, Djamgoz MB, Isom LL (2008) An emerging role for voltage-gated Na+ channels in cellular migration: regulation of central nervous system development and potentiation of invasive cancers. Neuroscientist 14:571–583

Brew HM, Hallows JL, Tempel BL (2003) Hyperexcitability and reduced low threshold potassium currents in auditory neurons of mice lacking the channel subunit Kv1. 1. J Physiol 548(1):1–20

Brini M, Carafoli E (2011) The plasma membrane Ca^{2+} ATPase and the plasma membrane sodium calcium exchanger cooperate in the regulation of cell calcium. Cold Spring Harb Perspect Biol 3(2):1–15

Brini M, Calì T, Ottolini D, Carafoli E (2013) The plasma membrane calcium pump in health and disease. FEBS J. 280(21):5385–5397

Brown DA, Passmore GM (2009) Neural KCNQ (Kv7) channels. Br J Pharmacol 156(8):1185–1195

Burganos VN (2017) Modeling and simulation of membrane structure and transport properties, reference module in chemistry. Mol Sci Chem Eng. ISBN 9780124095472

Cao G, Zhang M, Miao J, Li W, Wang J, Lu D, Xia J (2015) Effects of X-ray and carbon ion beam irradiation on membrane permeability and integrity in Saccharomyces cerevisiae cells. J Radiat Res 56(2):294–304

Catterall WA, Perez-Reyes E, Snutch TP, Striessnig J (2005) International union of pharmacology. XLVIII. Nomenclature and structure-function relationships of voltage-gated calcium channels. Pharmacol Rev 57:411–425
Cleal K, He L, Watson PD, Jones AT (2013) Endocytosis, intracellular traffic and fate of cell penetrating peptide based conjugates and nanoparticles. Curr Pharm Des 19(16): 2878–2894
Cooper GM (2000) The cell: a molecular approach, 2nd edn. Sinauer Associates, Sunderland
Cui C, Merritt R, Fu L, Pan Z (2017) Targeting calcium signaling in cancer therapy. Acta Pharmaceutica Sinica B 7(1):3–17R
Das A, Pushparaj C, Herreros J, Nager M, Vilella R, Portero M, Pamplona M-GX, Marti RM, Canti C (2013) T-type calcium channel blockers inhibit autophagy and promote apoptosis of malignant melanoma cells. Pigment Cell Melanoma Res 26:874–885
Dean M, Rzhetsky A, Allikmets R (2001) The human ATP-binding cassette (ABC) transporter superfamily. Genome Res 11(7):1156–1166
Delemotte L, Treptow W, Klein ML, Tarek M (2010) Effect of sensor domain mutations on the properties of voltage-gated ion channels: molecular dynamics studies of the potassium channel Kv1.2. Biophys J 99(9):L72–L74
Dziegielewska B, Brautigan DL, Larner JM, Dziegielewski J (2014) T-type Ca^{2+}channel inhibition induces p53-dependent cell growth arrest and apoptosis through activation of p38-MAPK in colon cancer cells. Mol Cancer Res 12:348–358
Eisenberg RS (2013) Ionic interactions in biological and physical systems: a variational treatment. Faraday Discuss 160:279–296
Fan Z, Kumon RE, Park J, Deng CX (2010) Intracellular delivery and calcium transients generated in sonoporation facilitated by microbubbles. J Control Release 142(1):31
Gavrilova-Ruch O, Schonherr K, Gessner G, Schonherr R, Klapperstuck T, Wohlrab W, Heinemann SH (2002) Effects of imipramine on ion channels and proliferation of IGR1 melanoma cells. J Membr Biol 188:137–149
Gehl J (2003) Electroporation: theory and methods, perspectives for drug delivery, gene therapy and research. Acta Physiol Scand 177(4):437–447
Gerbeau P, Amodeo G, Henzler T, Santoni V, Ripoche P, Maurel C (2002) The water permeability of Arabidopsis plasma membrane is regulated by divalent cations and pH. Plant J 1:71–81
Gillespie D, Eisenberg RS (2002) Physical descriptions of experimental selectivity measurements in ion channels. Eur Biophys J 31:454–466
Golowasch J, Nadim F (2014) Capacitance, membrane. Encyclopedia of computational neuroscience. Springer, Berlin\Heidelberg
Gomulkiewicz J, Bartoszkiewicz M, Miekisz S (2001) Some remarks on ion transport across excitable membranes. I. The stationary state. Curr Top Biophys 25:3–9
Gouaux E, MacKinnon R (2004) Principles of selective ion transport in channels and pumps. Science 310:1461–1465
Gutman GA, Chandy KG, Grissmer S, Lazdunski M, Mckinnon D, Pardo LA, Robertson GA, Rudy B, Sanguinetti MC, Stühmer W, Wang X (2005) International union of pharmacology. LIII. Nomenclature and molecular relationships of voltage-gated potassium channels. Pharmacol Rev 57(4):473–508
Hedfalk K, Törnroth-Horsefield S, Nyblom M, Johanson U, Kjellbom P, Neutze R (2006) Aquaporin gating. Curr Opin Struct Biol 16:447–456
Johnston J, Forsythe ID, Kopp-Scheinpflug C (2010) SYMPOSIUM REVIEW: Going native: voltage-gated potassium channels controlling neuronal excitability. J Physiol 588(17):3187–3200
Kaczorowski GJ, McManus OB, Priest BT, Garcia ML (2008) Ion channels as drug targets: the next GPCRs. J Gen Physiol 131(5):399–405
Kale VP, Amin SG, Pandey MK (2015) Pandey targeting ion channels for cancer therapy by repurposing the approved drugs. Biochim Biophys Acta 1848:2747–2755

Kotnik T, Kramar P, Pucihar G, Miklavcic D, Tarek M (2012) IEEE Electrical Insulation Magazine 28(5):14–23

Kruger LC, Isom LL (2016) Voltage-gated Na^+ channels: not just for conduction. Cold Spring Harb Perspect Biol 8(6):a029264

Kulbacka J, Pucek A, Kotulska M, Dubińska-Magiera M, Rossowska J, Rols MP, Wilk KA (2016a) Electroporation and lipid nanoparticles with cyanine IR-780 and flavonoids as efficient vectors to enhanced drug delivery in colon cancer. Bioelectrochemistry 110:19–31

Kulbacka J, Pucek A, Wilk KA, Dubińska-Magiera M, Rossowska J, Kulbacki M, Kotulska M (2016b) The effect of millisecond pulsed electric fields (msPEF) on intracellular drug transport with negatively charged large Nanocarriers made of solid lipid nanoparticles (SLN): in vitro study. J Membr Biol 249(5):645–661

Leonard GD, Fojo T, Bates SE (2003) The role of ABC transporters in clinical practice. Oncologist 8(5):411–424

Li J, Lin H (2011) Numerical simulation of molecular uptake via electroporation. Bioelectrochemistry 82(1):10–21

Li M, Xiong Z-G (2011) Ion channels as targets for cancer therapy. Int J Physiol Pathophysiol Pharmacol 3(2):156–166

Li Q, Shu Y (2014) Role of solute carriers in response to anticancer drugs. Molecular and Cellular Therapies 2(1):15

Lin L, Yee SW, Kim RB, Giacomini KM (2015) SLC transporters as therapeutic targets: emerging opportunities. Nature Reviews Drug Discovery 14(8):543–560

Linton KJ (2007) Structure and function of ABC transporters. Physiology (Bethesda) 22:122–130

Lodish H, Berk A, Zipursky SL, Matsudaira P, Baltimore D, Darnell J (2000) Molecular cell biology, 4th edn. W. H. Freeman, New York

McKeown L, Swanton L, Robinson P, Jones OT (2008) Surface expression and distribution of voltage gated potassium channels in neurons (review). Mol Membr Biol 25(4):332–343

Millward MJ, Cantwell BM, Munro NC, Robinson A, Corris PA, Harris AL (1993) Oral verapamil with chemotherapy for advanced non-small cell lung cancer: a randomized study. Br J Cancer 67:1031–1035

Min KA, Shin MC, Yu F, Yang M, David AE, Yang VC, Rosania GR (2013) Pulsed magnetic field improves the transport of iron oxide nanoparticles through cell barriers. ACS Nano 7(3):2161–2171

Mir LM, Orlowski S, Belehradek J, Paoletti C (1991) Electrochemotherapy potentiation of antitumour effect of bleomycin by local electric pulses. Eur J Cancer 27(1):68–72

Misonou H, Mohapatra DP, Menegola M, Trimmer JS (2005) Calcium-and metabolic state-dependent modulation of the voltage-dependent Kv2. 1 channel regulates neuronal excitability in response to ischemia. J Neurosci 25(48):11184–11193

Mohapatra DP, Misonou H, Sheng-Jun P, Held JE, Surmeier DJ, Trimmer JS (2009) Regulation of intrinsic excitability in hippocampal neurons by activity-dependent modulation of the KV2. 1 potassium channel. Channels 3(1):46–56

Muscella A, Calabriso N, Vetrugno C, Fanizzi FP, De Pascali SA, Storelli C (2011) The platinum (II) complex [Pt(O,O′-acac)(γ-acac)(DMS)] alters the intracellular calcium homeostasis in MCF-7 breast cancer cells. Biochem Pharmacol 81:91–103

Nejad SM, Hosseini H, Akiyama H, Tachibana K (2016) Reparable cell Sonoporation in suspension: Theranostic potential of microbubble. Theranostics 6(4):446–455

Nesin V, Bowman AM, Xiao S, Pakhomov AG (2012) Cell permeabilization and inhibition of voltage-gated Ca^{2+} and Na^+ channel currents by nanosecond pulsed electric fields. Bioelectromagnetics 33(5):394–404

O'Grady SM, Lee SY (2005) Molecular diversity and function of voltage-gated (Kv) potassium channels in epithelial cells. Int J Biochem Cell Biol 37:1578–1594

O'Malley HA, Isom LL (2015) Sodium channel b subunits: emerging targets in channelopathies. Annu Rev Physiol 77:481–504

Oh N, Park JH (2014) Endocytosis and exocytosis of nanoparticles in mammalian cells. Int J Nanomedicine 9(Suppl 1):51–63

Pakhomov AG, Kolb J, White J, Shevin R, Pakhomova ON, Schoenbach KS (2007a) Membrane effects of ultrashort (nanosecond) electric stimuli. Society for Neuroscience 37th Annual Meeting, San Diego, Nov 2–7 2007, Neuroscience meeting planner CD-ROM, Presentation No.: 317.14.

Pakhomov AG, Shevin R, White JA, Kolb JF, Pakhomova ON, Joshi RP, Schoenbach KH (2007b) Membrane permeabilization and cell damage by ultrashort electric field shocks. Arch Biochem Biophys. 465(1):109–118

Pardo LA, Stühmer W (2014) The roles of K(+) channels in cancer. Nat Rev Cancer 14(1):39–48

Patel F, Brackenbury WJ (2015) Dual roles of voltage-gated sodium channels in development and cancer. Int J Dev Biol 59(7–9):357–366

Pavlin M, Kandušer M, Reberšek M, Pucihar G, Hart FX, Magjarevićcacute R, Miklavčič D (2005) Effect of cell electroporation on the conductivity of a cell suspension. Biophys J 88(6):4378–4390

Perland E, Fredriksson R (2016) Classification Systems of Secondary Active Transporters. Trends Pharmacol Sci S0165-6147(16):30166–30163

Pitt WG, Husseini GA, Staples BJ (2004) Ultrasonic drug delivery – a general review. Expert Opin Drug Deliv 1(1):37–56

Qiao G, Duan W, Chatwin C, Sinclair A, Wang W (2010) Electrical properties of breast cancer cells from impedance measurement of cell suspensions. J Phys Conf Ser 224:012081

Rask-Andersen M, Masuram S, Fredriksson R, Schiöth HB (2013) Solute carriers as drug targets: Current use, clinical trials and prospective. Molecular Aspects of Medicine 34(2-3):702–710

Ray S, Kassan A, Busija AR, Rangamani P, Patel HH (2016) The plasma membrane as a capacitor for energy and metabolism. Am J Physiol Cell Physiol 310(3):181–192

Rim HK, Lee HW, Choi IS, Park JY, Choi HW, Choi JH, Cho YW, Lee JY, Lee KT (2012) T-type Ca^{2+} channel blocker, KYS05047 induces G1 phase cell cycle arrest by decreasing intracellular Ca^{2+} levels in human lung adenocarcinoma A549 cells. Bioorg Med Chem Lett 22:7123–7126

Rosazza C, Buntz A, Rieß T, Wöll D, Zumbusch A, Rols M-P (2013) Intracellular tracking of single-plasmid DNA particles after delivery by electroporation. Mol Ther 21(12):2217–2226

Runas KA, Malmstadt N (2015) Low levels of lipid oxidation radically increase the passive permeability of lipid bilayers. Soft Matter 11(3):499–505

Selwan EM, Finicle BT, Kim SM, Edinger AL (2016) Attacking the supply wagons to starve cancer cells to death. FEBS Lett 590(7):885–907

Shattock MJ, Ottolia M, Bers DM, Blaustein MP, Boguslavskyi A, Bossuyt J, Bridge JH, Chen-Izu Y, Clancy CE, Edwards A, Goldhaber J, Kaplan J, Lingrel JB, Pavlovic D, Philipson K, Sipido KR, Xie ZJ (2015) Na^+/Ca^{2+} exchange and Na^+/K^+-ATPase in the heart. J Physiol 593(6):1361–1382

Shigekawa M, Iwamoto T (2001) Cardiac Na^+-Ca^{2+} exchange. Circ Res 88:864–876

Shin JM, Munson K, Sachs G (2011) Gastric H+, K+-ATPase. Compr Physiol 1(4):2141–2153

Shukla S, Ohnuma S, Ambudkar SV (2011 May) Improving cancer chemotherapy with modulators of ABC drug transporters. Curr Drug Targets 12(5):621–630

Simms BA, Zamponi GW (2014) Neuronal voltage-gated calcium channels: structure, function, and dysfunction. Neuron 82:24–45

Steller L, Kreir M, Salzer R (2012) Natural and artificial ion channels for biosensing platforms. Anal Bioanal Chem 402(1):209–230

Strehler EE (2013) Plasma membrane calcium ATPases as novel candidates for therapeutic agent development. J Pharm Pharm Sci 16(2):190–206

Sugano K, Kansy M, Artursson P, Avdeef A, Bendels S, Di L, Ecker GF, Faller B, Fischer H, Gerebtzoff G, Lennernaes H, Senner F (2010) Coexistence of passive and carrier-mediated processes in drug transport. Nat Rev Drug Discov 9(8):597–614

Tamošiūnas M, Mir LM, Chen WS, Lihachev A, Venslauskas M, Šatkauskas S (2016) Intracellular delivery of bleomycin by combined application of electroporation and sonoporation in vitro. J Membr Biol 249(5):677–689

Tan Q, Ferrier GA, Chen BK, Wang C, Sun Y (2012) Quantification of the specific membrane capacitance of single cells using a microfluidic device and impedance spectroscopy measurement. Biomicrofluidics 6:034112

Taylor JM, Simpson RU (1992) Inhibition of cancer cell growth by calcium channel antagonists in the arrhythmic mouse. Cancer Res 52:2413–2418
Tombola F, Pathak MM, Isacoff EY (2006) How does voltage open an ion channel? Ann Rev Cell Dev Biol 22:23–52
Vasiliou V, Vasiliou K, Nebert DW (2009) Human ATP-binding cassette (ABC) transporter family. Hum Genomics 3(3):281–290
Wang L, Bhattacharjee A, Zuo Z, Hu F, Honkanen RE, Berggren PO, Li M (1999) A low voltage-activated Ca^{2+} current mediates cytokine-induced pancreatic beta-cell death. Endocrinology 140:1200–1204
Wang H, Zhang Y, Cao L, Han H, Wang J, Yang B, Nattel S, Wang Z (2002) HERG K+ channel, a regulator of tumor cell apoptosis and proliferation. Cancer Res 62:4843–4848
Waterbeemd H, Lennernäs H, Artursson P (2006) Drug bioavailability: estimation of solubility, permeability, absorption and bioavailability. John Wiley & Sons, Weinheim
Wezgowiec J, Derylo MB, Teissie J, Orio J, Rols M-P, Kulbacka J, Saczko J, Kotulska M (2013) Electric field-assisted delivery of photofrin to human breast carcinoma cells. J Membr Biol 246(10):725–735
Xiong Q, Gao Z, Wang W, Li M (2008) Activation of Kv7 (KCNQ) voltage-gated potassium channels by synthetic compounds. Trends Pharmacol Sci 9(2):99–107
Yang NJ, Hinner MJ (2015) Getting across the cell membrane: an verviewo for small molecules, peptides, and proteins. Methods in molecular biology (Clifton, N.J.), 1266, 29–53
Yellen G (2002) The voltage-gated potassium channels and their relatives. Nature 419(6902):35–42
Yukutake Y, Hirano Y, Suematsu M, Yasui M (2009) Rapid and reversible inhibition of aquaporin-4 by zinc. Biochemist 48:12059–12061
Zamponi GW (2016) Targeting voltage-gated calcium channels in neurological and psychiatric diseases. Nat Rev Drug Discov 15(1):19–34
Zamponi GW, Striessnig J, Koschak A, Dolphin AC (2015) The physiology, pathology, and pharmacology of voltage-gated calcium channels and their future therapeutic potential. Pharmacol Rev 67:821–870
Zeghimi A, Escoffre JM, Bouakaz A (2015) Role of endocytosis in sonoporation-mediated membrane permeabilization and uptake of small molecules: a electron microscopy study. Phys Biol 12:066007
Zhang S, Morris ME (2003) Effects of the flavonoids biochanin A, morin, phloretin, and silymarin on P-glycoprotein-mediated transport. J Pharmacol Exp Ther 304(3):1258–1267
Zhang WM, Zhou J, Ye QJ (2008) Endothelin-1 enhances proliferation of lung cancer cells by increasing intracellular free Ca^{2+}. Life Sci 82:764–771
Zhao Y, Zhao XT, Chen DY, Luo YN, Jiang M, Wei C, Long R, Yue WT, Wang JB, Chen J (2014) Tumor cell characterization and classification based on cellular specific membrane capacitance and cytoplasm conductivity. Biosens Bioelectron 57:245–253

Chapter 4
Cell Membrane Electropulsation: Chemical Analysis of Cell Membrane Modifications and Associated Transport Mechanisms

Antoine Azan, Florian Gaillìègue, Lluis M. Mir, and Marie Breton

Abstract The transport of substances across the cell membrane is complex because the main physiological role of the membrane is the control of the substances that would enter or exit the cells. Life would not be possible without this control. Cell electropulsation corresponds to the delivery of electric pulses to the cells and comprises cell electroporation and cell electropermeabilization. Cell electropulsation allows for the transport of non-permeant molecules across the membrane, bypassing the physiological limitations. In this chapter we discuss the changes occurring in the cell membrane during electroporation or electropermeabilization as they allow to understand which molecules can be transported as well as when and how their transport can occur. Electrophoretic or diffusive transports across the cell membrane can be distinguished. This understanding has a clear impact on the choice of the electrical parameters to be applied to the cells as well as on other aspects of the experimental protocols that have to be set to load the cells with non-permeant molecules.

4.1 Introduction

The cell membrane, also termed plasma membrane, highly regulates all the exchanges between the outside and the inside of the cell. In particular, the cell membrane is a barrier that prevents the unregulated penetration/leakage of molecules of vital importance for the cell physiology, such as all the substrates/products of the cell metabolism, ions, sugars, or amino acids. None of these molecules can cross the

A. Azan • F. Gaillìègue • L.M. Mir (✉) • M. Breton
Vectorology and Anticancer Therapies, UMR8203, Univ. Paris-Sud, CNRS, Gustave Roussy, Université Paris-Saclay, Villejuif F-94805, France
e-mail: Luis.mir@cnrs.fr

J. Kulbacka, S. Satkauskas (eds.), *Transport Across Natural and Modified Biological Membranes and its Implications in Physiology and Therapy*, Advances in Anatomy, Embryology and Cell Biology 227,
DOI 10.1007/978-3-319-56895-9_4

cell membrane by free diffusion. Other large hydrophilic molecules are also unable to cross the membrane by diffusion. These molecules and the small hydrophilic molecules that are not transported through channels or pumps are termed "non-permeant" molecules (Silve et al. 2012a); they can only reach certain cell compartments (but neither the cytosol nor the nucleosol) by endocytosis/exocytosis.

To allow the uptake of non-permeant molecules, cell membrane perturbations must be initiated that will transiently rupture the membrane impermeability. One of the most popular and practical way to reversibly permeabilize the cell membrane consists in the application of adequate electric pulses. The term electroporation was first introduced by Neumann and Katchalsky in 1972. The modelization of the phenomenon began almost as early and started with 1D approximations of the membrane (Litster 1975). The initial basis of analytical models was the Schwan equation:

$$\Delta\Psi = 3/2.E.R.\cos(\theta)$$

(the transmembrane potential $\Delta\Psi$ at a given point on the membrane is equal to the electric field amplitude E times the radius of the cell R times the cosine of the angular coordinate θ).

This equation implies that polarization in the cells occurs at the poles according to the electric field orientation and thus will in priority induce an electrical breakdown of the membrane in these regions.

There are two main mathematical approaches to model the pore formation: either static or dynamic. Older models focused on the static approach as in Pastushenko and Chizmadzhev (1982). The other approach, the dynamic one, considers the temporal evolution of the pore density (Krassowska and Filev 2007). This model links the number of pores to the voltage applied and then evaluates the evolution of the pore radii. The flaw of this model is that it is highly dependent on the mesh size. Using a fine mesh gives different results from a coarser one. Indeed, with the Krassowska model, small changes in initial non-measurable parameters can result in very different outcomes (Poignard et al. 2016). This can be a consequence of the fact that many initial variables are only approximations, because measuring them, for example, the conductivity and permittivity of the membrane or the cytosol, is possible (Wang et al. 2017) but difficult. Indeed, the cytosol is not just a solution of proteins and other biological compounds since the cytoskeleton and the vesicles occupy a large fraction of the cytoplasmic volume. To address this problem, a new model (the Leguèbe model) has been developed which relies on a lower number of variables. Moreover it considers two steps in the pulsing process: the generation of defects (possibly pores) at the poles of the cell membrane, followed by the diffusion of these defects in the cell membrane after the pulses delivery (Leguèbe et al. 2014). The Leguèbe model suggests a two-step approach of electropulsation. The membrane conductance fluctuates between the conductance at rest, the conductance during the pulse for a fully permeabilized membrane, and the conductance after the pulse which reflects the long-term effect of electroporation, also termed electropermeabilization.

Molecular dynamics (MD) have proven to be helpful to understand the interaction of cells with electric pulses. As far as the electroporation is concerned, MD allow for the study of the membrane structure at a nanoscale level. This computational method enables the simulation of the evolution of the atoms composing the membrane and its surroundings during up to tens of nanoseconds. If the force corresponding to the simulated electric field is high enough, pore generation can occur during the computational time. The process of pore generation is stochastic and simulations often require a high field amplitude. Since each individual molecule is considered in the calculations of the simulation, the region of interest is limited by the number of elements that the computer is able to handle simultaneously. Therefore, there are constraints in both the size of the sample considered and the duration of the simulation in order to avoid that running MD simulations are too computationally heavy. In particular, the nanosecond time range of this modeling is limited compared to the time during which the cell remains permeabilized. Nevertheless, in the case of nanosecond pulses, the predictions of the MD have been experimentally validated (Breton et al. 2012). More details of the results provided by these numerical approaches are reported in Chap. 1 of this volume.

Another relevant issue concerns the fact that the cell membrane is not the only membrane in the cell that may undergo modifications when the cell is submitted to an electric pulse. If nanosecond pulses are used, the membrane of internal organelles can also be porated (Schoenbach et al. 2001; de Menorval et al. 2016). Such an electroporation of the internal organelles of the cell may allow exchanges of molecules between different internal compartments of the cell, namely, between the cytosol and the endoplasmic reticulum or between the cytosol and the mitochondria. Therefore, after exposure to electric pulses, the complexity of the possible transports of molecules must be taken into account.

A further issue is that there is no consensus up to now about the term "electroporation." Electroporation refers to the generation of pores via pulsed electric fields (PEFs), while the term "electropermeabilization" refers to the generation of defects by PEFs. It is sometimes considered a more suitable term because even after the pore resealing, the cell can remain in a permeabilized state for periods of time that are orders of magnitude longer than the pulse duration. Recently the term "electropulsation" has been proposed to describe the action of submitting cells to electric fields.

In this chapter, we present recent findings that have improved the understanding of the interaction of electric pulses with cell membranes, and we address the mechanisms that may underlie the transport of molecules either during or after the pulse delivery.

4.2 Recent Experimental Data on Chemical Modifications of Membranes Submitted to Electric Pulses

It has been known for a long time that the application of electric pulses to cells induced the appearance of reactive oxygen species (ROS) in the cell medium (Bonnafous et al. 1999). It has also been reported that the addition of antioxidants could reduce the death rate of cells provoked by pulses (Gabriel and Teissié 1994). Finally, some groups used indirect measurements of phospholipid oxidation by absorbance studies to show that pulses could peroxidize membrane phospholipids (Benov et al. 1994). More recently, mass spectrometry has been used to directly investigate the chemical processes taking place during or after electropulsation at the level of membranes. The study was conducted on a well-known membrane model, the giant unilamellar vesicles (GUVs). These vesicles were constituted of either saturated or unsaturated phospholipids since these lipids are both present in all cell membranes. The mass spectra obtained clearly showed that electric pulses can induce the oxidation of the unsaturated membrane phospholipids. Whatever the duration of the pulses (millisecond, microsecond, or nanosecond pulses), unsaturated lipids were oxidized, while saturated lipids remained intact. The mass spectra clearly presented a classical oxidation pattern showing peaks corresponding to the addition of oxygen atoms as well as peaks of low masses corresponding to lipids that underwent chain breaks after oxidation. Mechanistic studies conducted on the same GUV model showed that the presence of light or dioxygen could be cofactors in the oxidation (M. Breton et al., submitted). More importantly, the concentration of ROS in the solution increased after pulsing only if unsaturated vesicles were present in the solution. Therefore, the electric pulses should not directly cause ROS generation and lipid peroxidation. It seems more likely that the main effect of the electric pulses is that they facilitate the peroxidation induced by the ROS already present in the solution before pulsing. Finally, the effects of the pulse parameters on the level of lipid oxidation follow the same trend as the effects of the pulse parameters on the level of cell permeability. Indeed, if the duration, the number, or the voltage of the pulses is increased, the level of oxidation is increased. This is the case for all kinds of pulses (millipulses, micropulses, nanopulses).

4.3 Cell Electropulsation Model

Our group proposed a new model which describes both the immediate effects of the pulses during their application to the cells as well as the chemical and physical consequences of the pulses on the cell membranes. One of the bases of this model is that when a very high electric field is applied on a membrane, non-permeant molecules such as water or ROS can enter the membrane first as water fingers and later as water pores (Tarek 2005). This model contains two steps: electroporation followed by electropermeabilization. Electroporation happens during and shortly after the application of the pulses. During this step, pores form, and therefore water and ROS can enter the lipophilic area of the membrane where the oxidable lipid chains are. The oxidation of the membrane phospholipids can take place. The membrane is then highly conductive and completely permeable. When the pulse is finished, the pores rapidly close in tens of nanoseconds. The electropermeabilization step initiates. The conductivity of the membrane sharply decreases, while the decrease of the membrane permeability is more progressive. However, since some oxidized lipids remain in the membrane until they are removed by membrane repair mechanisms, the conductivity and permeability of the membrane remain slightly higher than their basal level. During the electropermeabilization step, the membrane permeability will depend on the density of the oxidized lipids in the membrane. Right after the end of the pulse, the oxidized lipids are concentrated in the porated area. The density of oxidized lipids is high and the membrane is very permeable. Gradually, the oxidized lipids will diffuse laterally in the membrane. The density of oxidized lipids will decrease, and therefore the permeability of the membrane will decrease. A low level of membrane conductivity and permeability will remain until the complete removal of oxidized lipids from the membrane. Cells can renew their membrane lipids by the process of endocytosis/exocytosis which can last for several hours (Ullery et al. 2015). The membrane will hence remain slightly permeable and conductive for a few hours. This model is valid for all kinds of pulses ranging from nanosecond to millisecond duration. It presents different kinetics which are consistent with all the previous experimental studies that did not fit with the theories of electroporation or electropermeabilization. According to this model, it seems important to take into account both the physical and the chemical consequences of cell pulsation when studying the transport of molecules inside the cells by electrotransfer.

4.4 Study of the Interfacial Water Around Biological Samples Submitted to Electric Pulses

Many groups have studied the water/lipid interface based on vibrational spectroscopy techniques (Bonn et al. 2010; Gruenbaum and Skinner 2011; Nagata and Mukamel 2010; Nihonyanagi et al. 2013). It has been reported that the interfacial water molecules are highly organized due to strong hydrogen bonds close to lipid head groups (Lopez et al. 2004) and to preferential water orientation pointing toward the lipids tails (Chen et al. 2010). CARS microscopy has been used to investigate the orientation of water in the lipid/water interfaces of multilamellar vesicles (Cheng et al. 2003). Teissie and colleagues suggested that this interfacial water could be considered as the first barrier to overcome in order to allow molecules to cross the plasma membrane (Teissie 2007). Molecular dynamics simulations have revealed that the application of an intense electric field on a lipid membrane disorganizes the interfacial water by creating aqueous pores into the membrane (Tarek 2005). Water molecules play a key role in the initiation and the stabilization of these aqueous pores (Tieleman 2004; Ziegler and Vernier 2008). Due to the dipole moment of water, the water molecules are mainly oriented along the electric field, which contributes to stabilization of these aqueous pores (Tokman et al. 2013). The lifetime of these aqueous pores is still under debate in the community (Dehez et al. 2014; Pavlin and Miklavčič 2008; Ziegler and Vernier 2008). Nevertheless, it is commonly admitted that the aqueous pores are initiated and collapse within nanoseconds or tens of nanoseconds, respectively, after the beginning and the end of the pulsed electric fields (Delemotte and Tarek 2012). Due to the nanosecond time scale and the nanometer size scale of the electropores, obvious experimental proofs of the presence of the electropores created by the pulsed electric fields are still missing. In an experimental study, we focused on an indirect consequence of the delivery of pulsed electric fields on biological membranes by monitoring the interfacial water thanks to a unique Coherent anti-Stokes Raman scattering (CARS) microscope (Silve et al. 2012b). The illumination geometry specifically enhances the CARS signal at the interfaces. Thus, CARS spectra of water close to biological samples (DC-3F cells and GUVs) exposed or not to PEFs have been acquired. Two families of water, coined "interfacial water" and "interstitial water," are associated to two different vibrational wavenumber spectra bands. The 3000–3230 cm^{-1} and 3300–3450 cm^{-1} bands are known to be associated, respectively, to the interfacial water and the interstitial water (Gruenbaum and Skinner 2011). Our results show that the CARS intensity ratio between these two families is highly affected by the pulsed electric field delivery. The interstitial water becomes predominant in the water vibrational spectrum after the PEFs treatment. The differences in the spectra of GUVs before and after the pulsed electric field treatments are larger than those of the spectra of DC-3F cells. We associate this result with the absence of proteins in the membrane of GUVs. Indeed, the presence of proteins contributes to the reduction of the interfacial water/lipid signal. This study reports the first experimental proof of the effect of pulsed electric fields on the water and especially the water/lipid interface of biological samples. More experiments still have to be performed in order to improve the understanding of the underlying mechanisms of the electropermeabilization process such as the determination of a dose-effect relationship or the application of different types of electric pulses.

4.5 Biochemical Characterization of Live Cells Exposed to Pulsed Electric Fields

As reported in the previous section, mass spectrometry studies have demonstrated the chemical modification of phospholipids induced by the delivery of PEFs on GUVs. In order to assess the effect of pulsed electric fields on the biochemical composition of cells, confocal Raman microspectroscopy was used to acquire the Raman signature of live cells exposed or not to classical electric pulses parameters: 8 pulses, 100 μs duration, 1000 V/cm field magnitude, and 1 Hz repetition rate. Confocal Raman microspectroscopy is a label-free and noninvasive optical technique which provides detailed information about the molecular composition of the samples, especially about the proteins, lipids, and DNA contents of the cells. This instrumentation technique has reached a mature state since the discovery in 1928 of the physical underlying mechanism by Sir Raman (Raman 1928). It has been commonly used to characterize cells (Downes et al. 2011) and drug delivery systems (Smith et al. 2015) or to perform biomedical diagnoses (Kong et al. 2015). Major modifications in the Raman signatures of live cells were noticed when comparing the pulsed group to the control group when performing the acquisition in a cytoplasmic area (Fig. 4.1).

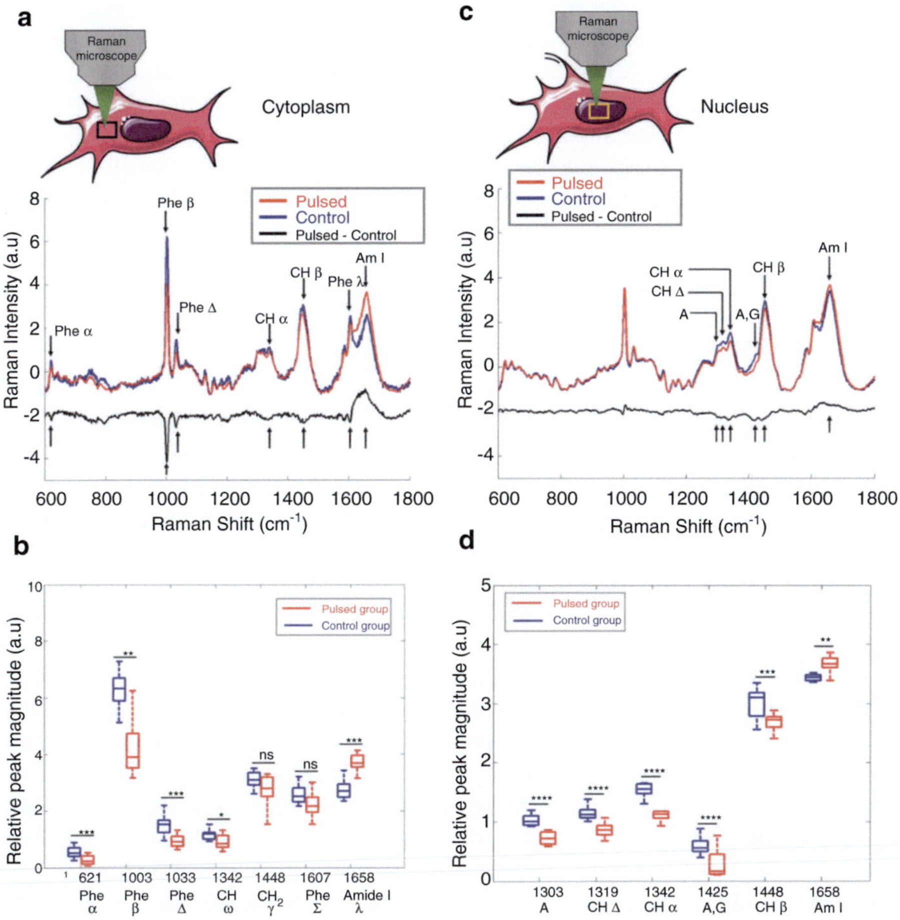

Fig. 4.1 Effect of the delivery of electric pulses on the Raman signatures of live cells at the cytoplasm (**a**, **b**) and the nucleus (**c**, **d**) locations. (**a**) Mean Raman signatures of control and pulsed cells acquired at the cytoplasm. The differential spectrum (pulsed minus control) is displayed with a vertical offset for more clarity. (**b**) Statistical analysis of the magnitude of seven critical bands. (**c**) Mean Raman signatures of control and pulsed cells acquired at the nucleus. The differential spectrum (pulsed minus control) is displayed with a vertical offset for more clarity. (**d**) Statistical analysis of the magnitude of six critical bands. The electric pulses parameters were fixed to eight pulses, 100 μs duration, 1000 V/cm field magnitude, and 1 Hz repetition rate. Statistics were conducted with a Student's t-test (*NS* non-statistically significant, *: *p*-value < 5%, **: *p*-value < 1%, ***: *p*-value < 0.1%, ****: *p*-value < 0.01%) (Adapted from Azan et al. (2017))

Especially, the vibrational modes of phenylalanine (621, 1003, 1033, and 1607 cm^{-1}), amide I (1658 cm^{-1}), and lipids (1448 cm^{-1}) were highly impacted by the delivery of pulsed electric fields (Fig. 4.1a, b) (Azan et al. 2017). Phenylalanine is a nonpolar and hydrophobic amino acid present in many transmembrane domains (Unterreitmeier et al. 2007). It has been demonstrated that the 1658 cm^{-1} Raman band is a biomarker of the secondary structure of proteins (Maiti et al. 2004). Our results experimentally confirm for the first time the effects of electric pulses on proteins (Azan et al. 2017) that were predicted by a numerical model showing the unfolding of proteins under an intense electric field (Cournia et al. 2015). When the Raman signatures of pulsed and non-pulsed cells were acquired at the nucleus location, vibrational models associated to DNA were also impacted by the pulsed electric fields (Fig. 4.1c, d). This result is in agreement with previous studies reporting that DNA is sensitive to the ROS generated in the culture medium by the delivery of electric pulses (Gabriel and Teissié 1994; Pakhomova et al. 2012; Wiseman and Halliwell 1996).

4.6 Transport Phenomena

Under the context of the theories and facts developed in the previous sections of this chapter, it is essential to distinguish the various transport possibilities that may occur across the membrane at the various steps of the electropulsation process. Here we will consider transport mechanisms under electropulsation conditions that preserve cell viability. Indeed, in the case of irreversible electroporation, when the membrane cannot fully recover its initial impermeability (one of the definitions of the "irreversible electroporation"), all molecules, sooner or later, will be able to freely cross the cell membrane.

Under normal conditions, in the absence of any physical or chemical perturbation, the membrane is impermeable to the non-permeant molecules, that is, to all the molecules that are not actively transported across the membrane and that cannot diffuse through the membrane (mainly through the lipid bilayer) because of their size and physicochemical characteristics. All the metabolically important small molecules (like sugars, amino acids, dipeptides, iron, osmotically important ions such as sodium or potassium, second messengers like calcium ions) do not freely cross the membrane: their transport is highly regulated. No large molecule can cross the plasma membrane, except the molecules that can generate pores in the membrane or affect the membrane structure (like the chains of the vegetal toxins responsible for the internalization of the catalytic chains of the same toxins or like the cell-penetrating peptides such as those derived from TAT or melittin) (Salomone et al. 2014).

When an electric pulse is delivered to cells, and as soon as the cell enters the first step of the membrane destabilization caused by the electric pulse, that is, the "electroporation" step, non-permeant molecules can start crossing the membrane (Fig. 4.2). They can cross either by diffusion through the aqueous pores created in the membrane or by electrophoresis. Indeed, because of the presence of the electric field, charged molecules will be efficiently pulled through the membrane. For highly charged molecules or, better to say, for molecules possessing a high ratio of charges per mass, this transport can be very efficient even for ultrashort pulses as demonstrated by Breton et al. (2012). In 10 ns, one pulse of 10 ns and 5.8 kV/cm was sufficient to permeabilize the membrane of GUVs and to introduce siRNAs inside the GUVs. Limited diffusive transport also occurs, but at a low level, for several reasons.

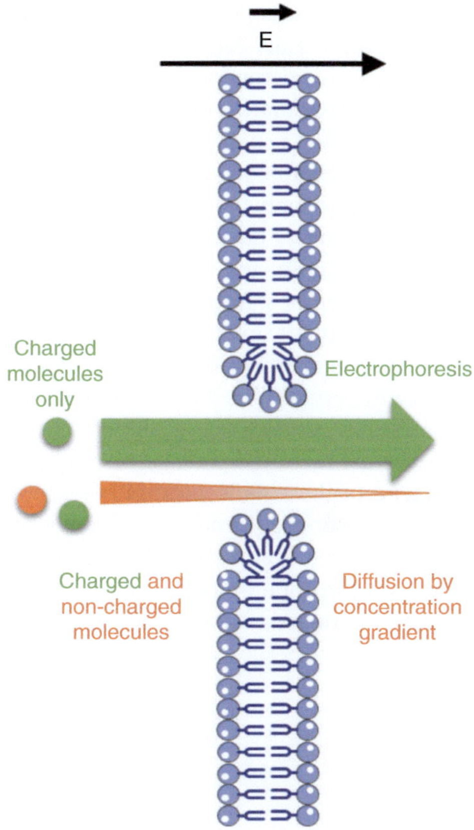

Fig. 4.2 Scheme presenting the transport of molecules during the electroporation step. The electric field is present. The duration of this step is directly related to the duration of the electric pulse, and so it ranges from a few nanoseconds to milliseconds. Charged molecules are mostly pulled through the membrane by electrophoresis. Limited diffusive transport also occurs for both charged and non-charged molecules

The first one is that the duration of the pulses, and therefore the half-life of the electropores, is always very short to preserve the cell viability. The second one is that this diffusion will only occur through the electropores, that is, through a small area of the cell surface, which will restrict the flux of molecules across the membrane. The third one is that unless there is a huge concentration gradient between the outside and the inside of the cell, the number of molecules transported through a small area during an extremely short time (Fick's law) will be very small and thus not easily detectable. Nevertheless, this transport exists and could be shown using GUVs prepared with a non-oxidable lipid, for example, the DSPC that is a fully saturated phospholipid. Because GUVs are just a phantom of a cell, they can be pulsed many times without the restrictions imposed by viability preservation. Because the lipids of these DSPC GUVs cannot be oxidized, even if pores form during the pulse delivery, the membrane recovers its full integrity when the pores close. DSPC GUVs prepared in 240 mM sucrose (sucrose inside the GUVs) and diluted in 260 mM glucose (glucose outside the GUVs) show a strong optical contrast. Two minutes after one single pulse, or a very limited number of pulses, no change in this contrast is observed (while GUVs prepared with an unsaturated oxidable lipid will already lose this contrast, see next paragraph). However, after the delivery of a high number of long pulses (e.g., of 5 ms duration), a loss in contrast is detected: it reflects the exchange of glucose and sucrose across the electropores generated during the pulse delivery. The cumulated duration of the pulse and the large concentration gradients of the two sugars allow for a sufficient mass transport of the sugar molecules across the membrane, and therefore the contrast loss, even if sugars are neutral, non-charged molecules. Thus, not only the electrophoretic transport can occur through the pores during the pulses delivery but also the diffusive transport.

After the pulse, and after the rapid closure of the electropores, the cells remain permeabilized for some time, in the seconds to minutes range (Fig. 4.3). No electrophoretic transport driven by an external electric field can occur since the electric pulses have ended. The contribution of a transmembrane potential difference across the electropermeabilized membrane has been evoked by Dr. T. Vernier and his colleagues and is under study: the electric field across the membrane could indeed favor or not favor the passage of charged species, as a function of the sign of the charge(s) of these molecules. However, with respect to the "electroporation" step, the "electropermeabilized" state is characterized by the massive transport of charged or non-charged non-permeant molecules across the membrane. One of the main reasons is the duration of the "electropermeabilized" state: seconds or minutes, as already mentioned. These durations are several orders of magnitude longer than the

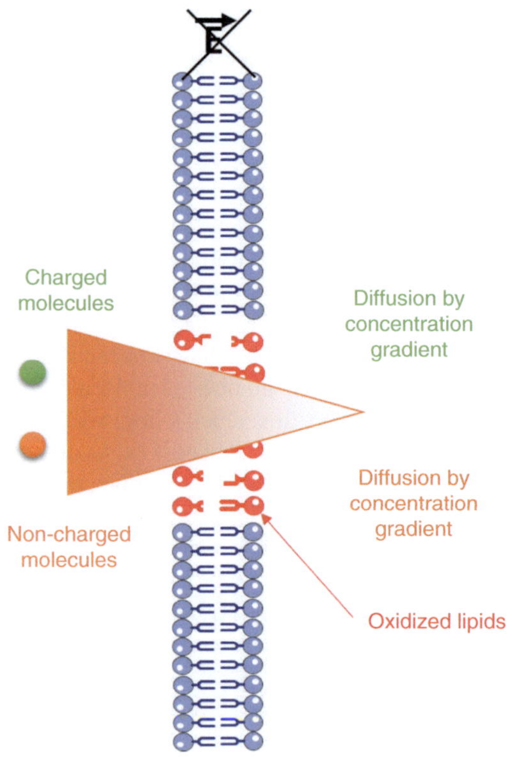

Fig. 4.3 Scheme presenting the transport of molecules during the electropermeabilization step. There is no electric field. This step has a long duration ranging from seconds to hours. The membrane is leaky because it contains oxidized phospholipids. Charged and non-charged molecules can be massively transported by diffusion if a concentration gradient is present

pulse duration, even in the case of the longest pulses, the msPEF. The duration of the permeabilized state is the consequence of the lipid oxidation, since it is well known that a membrane containing oxidized lipids (Wong-Ekkabut et al. 2007; Vernier et al. 2009; Rems et al. 2016) is a "leaky" membrane. The coefficient of diffusion through an oxidized membrane can be orders of magnitude larger than the coefficient of diffusion through a native, nonoxidized membrane (Rems et al. 2016). This explains why the optical contrast of a GUV prepared using an unsaturated lipid is lost 2 min after the delivery of a single pulse or very small number of pulses: sugars could cross the membrane during the 2 min, a time extremely long with respect to the pulse duration. This diffusive transport is, of course, not constant during the duration of the electropermeabilized state. There are probably two reasons why very large molecules have to be present in the pulsing medium at the time of the pulse delivery: (1) in the case of highly charged molecules, such as long nucleic acid that are too large to move through diffusive processes, to benefit from the electrophoretic transport and to at least initiate the contact with the membrane; (2) in the case of large non-charged molecules such as large proteins, to find regions with a very large density of oxidized lipids, therefore, to find regions with very large diffusive capacities. It is important to recall that DNA fragments of up to 150 kbp or proteins of 150 kDa, which were complete antibodies (Bobinnec et al. 1998), have been transported inside the cells by electropulsation. According to our model, and in particular according to the lateral diffusion of the lipids that will "dilute" the oxidized lipids generated in the areas where electropores were formed, the density of the modified lipids and hence the diffusion capabilities will progressively decrease, which will progressively restrict the diffusion of the molecules as a function of their size. The kinetics of such a progressive restriction in the molecule transport have been analyzed in the case of the anticancer drug bleomycin (Silve et al. 2012a). In another respect, the total number of molecules transported (independently of the external concentration and thus of the gradient of the concentration across the membrane) is directly related to the size of the molecules transported, under identical pulsation conditions and using the same cells to evaluate these transports (Mir 2008).

In conclusion, the properties of the molecules transport across the electropulsed membranes are in agreement with the membrane impermeability rupture model. During the electroporation phase, the electrophoretic transport of non-permeant molecules across the electropulsed membrane is more effective than the diffusive transport, while in the electropermeabilization phase, the diffusive transport will be the most efficient one.

References

Azan A, Untereiner V, Gobinet C, Sockalingum GD, Breton M, Piot O, Mir LM (2017) Demonstration of the protein involvement in cell electropermeabilization using confocal Raman microspectroscopy. Sci Rep 7:297–306

Benov LC, Antonov PA, Ribarov SR (1994) Oxidative damage of the membrane lipids after electroporation. Gen Physiol Biophys 13:85–97

Bobinnec Y, Khodjakov A, Mir LM, Rieder CL, Eddé B, Bornens M (1998) Centriole disassembly in vivo and its effect on centrosome structure and function in vertebrate cells. J Cell Biol 143:1575–1589

Bonn M, Bakker HJ, Ghosh A, Yamamoto S, Sovago M, Campen RK (2010) Structural inhomogeneity of interfacial water at lipid monolayers revealed by surface-specific vibrational pump–probe spectroscopy. J Am Chem Soc 132:14971–14978

Bonnafous P, Vernhes M, Teissié J, Gabriel B (1999) The generation of reactive-oxygen species associated with long-lasting pulse-induced electropermeabilisation of mammalian cells is based on a non-destructive alteration of the plasma membrane. Biochim Biophys Acta 1461:123–134

Breton M, Delemotte L, Silve A, Mir LM, Tarek M (2012) Transport of siRNA through lipid membranes driven by nanosecond electric pulses: an experimental and computational study. J Am Chem Soc 134:13938–13941

Chen X, Hua W, Huang Z, Allen HC (2010) Interfacial water structure associated with phospholipid membranes studied by phase-sensitive vibrational sum frequency generation spectroscopy. J Am Chem Soc 132:11336–11342

Cheng J-X, Pautot S, Weitz DA, Xie XS (2003) Ordering of water molecules between phospholipid bilayers visualized by coherent anti-stokes Raman scattering microscopy. Proc Natl Acad Sci U S A 100:9826–9830

Cournia Z, Allen TW, Andricioaei I, Antonny B, Baum D, Brannigan G, Buchete N-V, Deckman JT, Delemotte L, Del Val C et al (2015) Membrane protein structure, function, and dynamics: a perspective from experiments and theory. J Membr Biol 248:611–640

Dehez F, Delemotte L, Kramar P, Miklavčič D, Tarek M (2014) Evidence of conducting hydrophobic nanopores across membranes in response to an electric field. J Phys Chem C 118:6752–6757

Delemotte L, Tarek M (2012) Molecular dynamics simulations of lipid membrane electroporation. J Membr Biol 245:531–543

Downes A, Mouras R, Bagnaninchi P, Elfick A (2011) Raman spectroscopy and CARS microscopy of stem cells and their derivatives. J Raman Spectrosc 42:1864–1870

Gabriel B, Teissié J (1994) Generation of reactive-oxygen species induced by electropermeabilization of Chinese hamster ovary cells and their consequence on cell viability. Eur J Biochem 223:25–33

Gruenbaum SM, Skinner JL (2011) Vibrational spectroscopy of water in hydrated lipid multibilayers. I. Infrared spectra and ultrafast pump-probe observables. J Chem Phys 135:34–36

Kong K, Kendall C, Stone N, Notingher I (2015) Raman spectroscopy for medical diagnostics – from in-vitro biofluid assays to in-vivo cancer detection. Adv Drug Deliv Rev 89:121–134

Krassowska W, Filev PD (2007) Modeling electroporation in a single cell. Biophys J 92:404–417

Leguèbe M, Silve A, Mir LM, Poignard C (2014) Conducting and permeable states of cell membrane submitted to high voltage pulses: mathematical and numerical studies validated by the experiments. J Theor Biol 360:83–94

Litster JD (1975) Stability of lipid bilayers and red blood cell membranes. Phys Lett A 53:193–194

Lopez CF, Lopez CF, Nielsen SO, Klein ML, Moore PB (2004) Hydrogen bonding structure and dynamics of water at the dimyristoylphosphatidylcholine lipid bilayer surface from a molecular dynamics simulation. J Phys Chem B 108:6603–6610

Maiti NC, Apetri MM, Zagorski MG, Carey PR, Anderson VE (2004) Raman spectroscopic characterization of secondary structure in natively unfolded proteins: alpha-synuclein. J Am Chem Soc 126:2399–2408

de Menorval M-A, Andre FM, Silve A, Dalmay C, Français O, Le Pioufle B, Mir LM (2016) Electric pulses: a flexible tool to manipulate cytosolic calcium concentrations and generate spontaneous-like calcium oscillations in mesenchymal stem cells. Sci Rep 6:32331

Mir LM (2008) Application of electroporation gene therapy: past, current, and future. Methods Mol Biol 423:3–17

Nagata Y, Mukamel S (2010) Vibrational sum-frequency generation spectroscopy at the water/lipid interface: molecular dynamics simulation study. J Am Chem Soc 132:6434–6442

Neumann E, Katchalsky A (1972) Long-lived conformation changes induced by electric impulses in biopolymers. Proc Natl Acad Sci U S A 69:993–997

Nihonyanagi S, Mondal JA, Yamaguchi S, Tahara T (2013) Structure and dynamics of interfacial water studied by heterodyne-detected vibrational sum-frequency generation. Annu Rev Phys Chem 64:579–603

Pakhomova ON, Khorokhorina VA, Bowman AM, Rodaitė-Riševičienė R, Saulis G, Xiao S, Pakhomov AG (2012) Oxidative effects of nanosecond pulsed electric field exposure in cells and cell-free media. Arch Biochem Biophys 527:55–64

Pastushenko VF, Chizmadzhev YA (1982) Stabilization of conducting pores in BLM by electric current. Gen Physiol Biophys 1:43–52

Pavlin M, Miklavčič D (2008) Theoretical and experimental analysis of conductivity, ion diffusion and molecular transport during cell electroporation – relation between short-lived and long-lived pores. Bioelectrochemistry 74:38–46

Poignard C, Silve A, Wegner L (2016) Different approaches used in modeling of cell membrane electroporation. Springer, New York

Raman CV (1928) A new radiation. Indian J Phys 2:387–398

Rems L, Tarek M, Casciola M, Miklavčič D (2016) Properties of lipid electropores II: comparison of continuum-level modeling of pore conductance to molecular dynamics simulations. Bioelectrochemistry 112:112–124

Salomone F, Breton M, Leray I, Cardarelli F, Boccardi C, Bonhenry D, Tarek M, Mir LM, Beltram F (2014) High-yield nontoxic gene transfer through conjugation of the CM $_{18}$ -tat $_{11}$ chimeric peptide with nanosecond electric pulses. Mol Pharm 11:2466–2474

Schoenbach KH, Beebe SJ, Buescher ES (2001) Intracellular effect of ultrashort electrical pulses. Bioelectromagnetics 22:440–448

Silve A, Leray I, Mir LM (2012a) Demonstration of cell membrane permeabilization to medium-sized molecules caused by a single 10 ns electric pulse. Bioelectrochemistry 87:260–264

Silve A, Dorval N, Schmid T, Mir LM, Attal-Tretout B (2012b) A wide-field arrangement for single-shot CARS imaging of living cells. J Raman Spectrosc 43:644–650

Smith GPS, McGoverin CM, Fraser SJ, Gordon KC (2015) Raman imaging of drug delivery systems. Adv Drug Deliv Rev 89:21–41

Tarek M (2005) Membrane electroporation: a molecular dynamics simulation. Biophys J 88:4045–4053

Teissie J (2007) Biophysical effects of electric fields on membrane water interfaces: a mini review. Eur Biophys J 36:967–972

Tieleman DP (2004) The molecular basis of electroporation. BMC Biochem 5:10

Tokman M, Lee JH, Levine Z a, Ho M-CC, Colvin ME, Vernier PT (2013) Electric field-driven water dipoles: nanoscale architecture of electroporation. PLoS One 8:e61111

Ullery JC, Tarango M, Roth CC, Ibey BL (2015) Activation of autophagy in response to nanosecond pulsed electric field exposure. Biochem Biophys Res Commun 458:411–417

Unterreitmeier S, Fuchs A, Schäffler T, Heym RG, Frishman D, Langosch D (2007) Phenylalanine promotes interaction of transmembrane domains via GxxxG motifs. J Mol Biol 374:705–718

Vernier PT, Levine ZA, Wu Y-H, Joubert V, Ziegler MJ, Mir LM, Tieleman DP (2009) Electroporating fields target oxidatively damaged areas in the cell membrane. PLoS One 4:e7966

Wang K, Zhao Y, Chen D, Fan B, Lu Y, Chen L, Long R, Wang J, Chen J (2017) Specific membrane capacitance, cytoplasm conductivity and instantaneous Young's modulus of single tumour cells. Sci Data 4:170015

Wiseman H, Halliwell B (1996) Damage to DNA by reactive oxygen and nitrogen species: role in inflammatory disease and progression to cancer. Biochem J 313(Pt 1):17–29

Wong-Ekkabut J, Xu ZT, Triampo W, Tang IM, Tieleman DP, Monticelli L (2007) Effect of lipid peroxidation on the properties of lipid bilayers: a molecular dynamics study. Biophys J 93:4225–4236

Ziegler MJ, Vernier PT (2008) Interface water dynamics and porating electric fields for phospholipid bilayers. J Phys Chem B 112:13588–13596

Chapter 5
Physical Methods for Drug and Gene Delivery Through the Cell Plasma Membrane

Milda Jakutavičiūtė, Paulius Ruzgys, Mindaugas Tamošiūnas, Martynas Maciulevičius, and Saulius Šatkauskas

Abstract The cell membrane represents a major barrier for efficient delivery of exogenous molecules, either pharmaceuticals or genetic material, under both in vitro and in vivo conditions. The number of methods employed to attempt safe, efficient, and local drug and gene delivery has increased during the recent years. One method for membrane permeabilization, electroporation, has already been translated to clinical practice for localized anticancer drug delivery and is termed "electrochemotherapy". Clinical trials for gene delivery using electroporation as well as drug delivery using another cell permeabilization method, sonoporation, are also underway. This review focuses on these two methods, including their fundamental principles and state-of-the-art applications. Other techniques, such as micro-injection, magnetoporation, photoporation, electrospray, and hydrodynamic and ballistic gene delivery, are also discussed.

5.1 Introduction

Targeted and controlled drug and gene delivery into cells and tissues remains a major topic for many research groups. Various drug and gene delivery methods are intensely investigated in order to develop efficient treatment methods for different congenital or acquired diseases. As compared to viral vectors, nonviral gene delivery methods have the advantage of easier preparation, possibility to transfer larger genes, and less safety risks, but they suffer from lower transfection efficiency and poor transgene expression. For more detailed comparison between nonviral and viral gene delivery

M. Jakutavičiūtė • P. Ruzgys • M. Tamošiūnas • M. Maciulevičius • S. Šatkauskas (✉)
Biophysical Research Group, Faculty of natural Sciences, Vytautas Magnus University, Vileikos 8, Kaunas LT-44404, Lithuania
e-mail: saulius.satkauskas@vdu.lt

J. Kulbacka, S. Satkauskas (eds.), *Transport Across Natural and Modified Biological Membranes and its Implications in Physiology and Therapy*, Advances in Anatomy, Embryology and Cell Biology 227,
DOI 10.1007/978-3-319-56895-9_5

methods, see Boulaiz et al. (2005) and Wang et al. (2013). To overcome the drawbacks of nonviral gene therapy, it is important to develop new gene delivery methods and to improve the existing ones. Similarly, targeted and controlled drug delivery systems can improve the drug stability and increase the effective concentration in target tissue while reducing the amount of the drug used for treatment and, thus, minimizing the side effects associated with conventional routes of administration (Balmayor et al. 2011). Because the development of new drugs is expensive and time-consuming, development of new delivery methods for existing drugs is a promising alternative (Tiwari et al. 2012). Some of these new drug and gene delivery methods have already been tested in clinical trials and are currently in the process of further optimization, while others are still in preclinical development phases.

The aim of this review is to provide a comprehensive update about various techniques and methods applied to deliver exogenous molecules into cells and tissues. Special focus is given to the two most developed physical delivery methods, namely, cell electroporation and sonoporation.

5.2 Microinjection

The first application of a microinjection method for the delivery of nucleic acids directly to the inside of a cell was reported more than 40 years ago (Das et al. 2015). The use of glass micropipettes with a narrow (less than 0.5 μm) tip enabled genetic material injection to cells with survival rates of up to 100% (Celis 1977). However, the microinjection is technically demanding and limited to treatment of only a few cells during one procedure.

There are three main applications for the microinjection technology: blastocyst microinjection, sperm-mediated gene transfer (SMGT), and intracytoplasmic sperm injection-mediated gene transfer (ICSI-MGT) (Fig. 5.1a).

5.3 DNA Bombardment: “Gene Gun”

DNA-coated particle bombardment or ballistic DNA delivery, nowadays known as gene gun, has originally been developed for gene delivery to plant tissue more than 30 years ago. The method is based on the use of a pressurized inert gas to induce an acceleration of plasmid-coated particles, which are then directed to bombard the tissue of interest (Klein et al. 1987) (Fig. 5.1b). Accelerated particles, usually gold particles in size of approximately 1 μm, penetrate through the plasma membrane into the cytosol, where plasmid DNA is later processed by the transcription machinery of the cell. Although the technology allows treatment of large surfaces, the main disadvantages are related to limited, few-millimeter-deep penetration of plasmid DNA, thus only allowing transfection of the surface of the target tissue (Williams et al. 1991). On the other hand, the ballistic DNA delivery

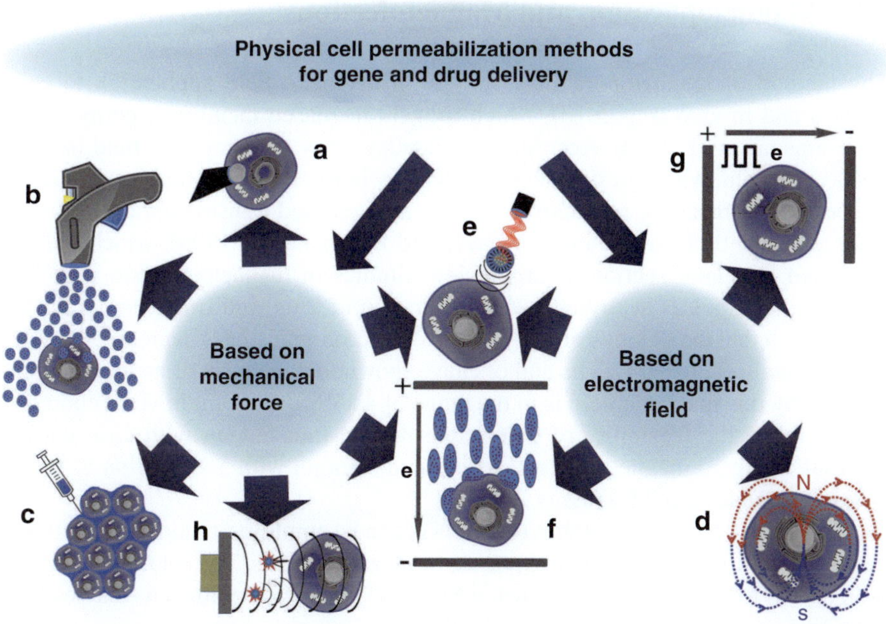

Fig. 5.1 Representation of different physical cell membrane permeabilization methods based on mechanical force and/or electromagnetic field

to cells is local and chemically nontoxic and enables delivery of large-size plasmid DNA (Lin et al. 2000). Ballistic DNA delivery has been applied for genetic vaccination (Hooper et al. 1999), suicide gene therapy (Klatzmann et al. 1998), and immunomodulation (Irvine et al. 1996).

5.4 Hydrodynamic Gene Delivery

Hydrodynamics-based gene delivery uses injection of large volumes of liquid to tissues to induce hydrodynamic pressure and intracellular gene transfer (Liu et al. 1999) (Fig. 5.1c). The most successful application of hydrodynamic gene delivery is the induction of hepatocyte transfection in rodents (Suda and Liu 2007). In this procedure, a volume (equivalent to 8–10% of body weight) of DNA containing physiological solution is injected into the tail vein within 5–7 s. It is hypothesized that plasmid DNA moves into the cytoplasm through hydrodynamically induced pores in the membrane (Kobayashi et al. 2004). This procedure is reported to achieve transfection of up to 40% of the hepatocytes (Zhang et al. 2004). Hydrodynamic gene delivery was not limited to hepatocytes but has been successfully performed on various animal models and various target tissues including the liver, hepatocellular carcinoma, muscle, kidney, spleen, pancreas, and brain (Bonamassa et al. 2011).

5.5 Magnetoporation and Magnetofection

The method of magnetoporation is based on magnetic field application to cells in vitro and in vivo (Fig. 5.1d), which triggers increased membrane permeability (Lakshmanan et al. 2014). According to Faraday's law, the magnetic field induces an electric field, which in turn induces changes in the transmembrane potential of the cell. After reaching a certain threshold, an increased transmembrane potential leads to the cell electroporation (Weaver and Chizmadzhev 1996), which will be discussed in a later section of this review. Studies in magnetoporation with 200 magnetic field pulses at strength of 7 T, duration of 2 μs, and frequency of 35 Hz have shown that the colony-forming ability of microorganisms is decreased by 90% (Markovskaja et al. 2014). The authors have shown that the peak of an electric field induced by this magnetic field is in the range of 4 kV/m. A recent in vivo report has demonstrated the feasibility of guinea pig dermal transfection with GFP coding plasmid, using 50 magnetic pulses of 4 T field strength with 5 s duration at 10 Hz frequency (Kardos and Rabussay 2012).

Multiwalled carbon nanotubes (MWCNT) can be employed to induce membrane permeabilization of cells under exposure to magnetic fields. The rotating magnetic field causes MWCNT to rotate, causing MWCNT interaction with cells that leads to membrane permeabilization (Liu et al. 2012).

Similarly to magnetoporation, the method of magnetofection also relies on magnetic fields to cause DNA delivery to cells (Fig. 5.1d). In this case, core-shell magnetic nanoparticles in the mixture with DNA can be magnetically guided to enhance nucleic acid delivery to cells (Fraites et al. 2000). Magnetic nanoparticles have been shown to increase the efficiency of viral delivery and lipofection: administering a mixture of these magnetic nanoparticles with viruses or polyethyleneimine (PEI) and plasmids to cells in vitro or in vivo and exposing cells to magnetic field with intensity from 0.6 to 2.1 T leads to enhanced transfection (Chen et al. 2006). Magnetic field induces an interaction between magnetic nanoparticles and the cell membrane which activates endocytotic pathways and allows plasmid DNA entry into the cell (Namgung et al. 2010).

Moreover, activated endocytotic pathways can facilitate viral vector infection even in the absence of virus receptors. This enables controlled viral transfections even in therapeutic applications (Tresilwised et al. 2010).

Magnetic field application for DNA delivery via magnetoporation or magnetofection is a promising emerging tool for gene transfer. However, the possible applications of these methods are still poorly documented.

5.6 Photoporation (Optoporation)

Photoporation or optoporation is a process in which a highly focused laser beam is used as an "optical" microbeam to facilitate uptake of exogenous DNA or other molecules into cells (Yao et al. 2008) (Fig. 5.1e). The first reported study showed

DNA transfer to cultured NRK cells using 355 nm laser at 1 mJ intensity, 5-ns pulse duration in spot size of 0.5 μm (Tsukakoshi et al. 1984). After this pioneering study, results showing the transfection of various eukaryotic cell lines with different wavelength lasers, 355 nm (Tao et al. 1987), 405 nm (Paterson et al. 2005), and 1065 nm (Mohanty et al. 2003), and also femtosecond lasers (Zeira et al. 2003) have been published. In all of these studies, genetic material transfer to cells has been achieved when pulsed laser irradiation perforated holes in cell membrane. The membrane poration is related to laser-induced stress waves (LISW). LISW are thought to be generated by high-intensity laser irradiation-mediated ionization of the medium and formation of the plasma. The generated plasma expands rapidly due to high temperature and pressure, and a high pressure front in the range of kilobars is created (Vogel and Busch 1994; Noack and Vogel 1995). Plasma can also cause the formation of bubbles that ultimately collapse and generate secondary shock waves (Doukas et al. 1996). An alternative hypothesis relates permeabilization to the thermal expansion of the medium that propagates as a stress wave (Doukas et al. 1996) (Fig. 5.1e).

Cell targeting with light-absorbing particles (CTLAP) is another method of molecular delivery to cells that utilizes the laser irradiation, similar to photoporation. In this case, cells are first coupled with light-absorbing particles and later placed in a medium containing plasmid DNA and irradiated with laser, which induces cell permeabilization (Pitsillides et al. 2003; Umebayashi et al. 2003) by mechanisms similar to LISW (Doukas et al. 1996).

5.7 Electrospray

Electrospray technology is emerging as another method for intracellular drug and gene delivery (Chen et al. 2000). Electrospray mechanism is based on generation of fine droplets of suspension that bombard cells of interest (Hradetzky et al. 2012). The generation of droplets is based on Coulomb's law. Electrospray causes successful cell transfection with droplets ranging from 100 to 1000 μm in size (Okubo et al. 2008). The key mechanism for electrospray-based cell membrane permeabilization is thought to be based on shear forces that are induced for a brief moment (10^{-5}–10^{-6} s) due to the contact of high-velocity droplets and the cell membrane (Ikemoto et al. 2012) (Fig. 5.1f). However, the potential as well as the mechanisms of electrospray-mediated molecule transfer has to be further investigated.

5.8 Electroporation

The phenomenon of electroporation can be traced back to the works of Neumann and colleagues in the early seventies (Neumann and Rosenheck 1972). This phenomenon relies on the induction of a change in the transmembrane potential of the cell upon its exposure to an electric field (Fig. 5.1g). As the induced transmembrane potential reaches a critical value, the local plasma membrane permeability to membrane-impermeant molecules increases (Hibino et al. 1991), presumably due to formation of so-called electropores. The critical value of transmembrane voltage needed to induce electroporation differs between different cell types and is estimated to be between 0.4 and 1 V (Hibino et al. 1993; Teissié and Rols 1993). However, the electric field-induced change in the transmembrane voltage is not distributed evenly along the cell surface. The change in the transmembrane potential induced at a particular point on cell surface is calculated via Formula (5.1) (Bernhardt and Pauly 1973):

$$\Delta\psi_i = f\, g(\lambda) r \cos(\theta) \tag{5.1}$$

Since cos (θ) in the formula is the cosine of angle between the direction of electric field and the point on the cell membrane where the induced transmembrane potential is measured, the poles of the cell which face the electrodes receive the highest change in the transmembrane potential and are the most electroporated, whereas the poles of cell membrane which are located at a 90° angle to the normal of electric field are not affected. Also, due to the resting transmembrane potential, the effect on the electrode-facing poles is not even, because the anode-facing pole is depolarized and the cathode-facing pole is hyperpolarized (Tekle et al. 1994). More detailed explanations of the phenomenon of the electroporation can be found in Chaps. 2, 3, and 7 of this book.

As long as the electric field is sufficient to electroporate the plasma membrane, small molecule (<4 kDa) electrotransfer will occur (Wolf et al. 1994; Gabriel and Teissié 1998). Small molecules pass through electroporated plasma membrane mainly due to simple diffusion (Gabriel and Teissié 1999). For charge-carrying molecules, the transport is facilitated by electrophoretic forces throughout the duration of the electric pulse (Pucihar et al. 2008), with simple diffusion carrying on afterward. The efficiency of small molecule electrotransfer depends on the parameters of the small molecule (i.e., size and charge) (Venslauskas et al. 2010), the electric pulse parameters (amplitude, duration, number, frequency) (Rols and Teissié 1990; Wolf et al. 1994; Canatella et al. 2001), and the parameters of the external medium (Djuzenova et al. 1996; Sadik et al. 2013).

Visualization experiments have shown that the transfer of small molecules begins milliseconds after the application of an over-threshold electric pulse (Gabriel and Teissié 1997, 1998, 1999; Escoffre et al. 2011). With low electric fields, the transfer only occurs from the anode-facing side (Gabriel and Teissié 1997, 1998, 1999). With higher electric fields, it happens from both electrode-facing poles, with higher transfer on the anode-facing side (Gabriel and Teissié 1997, 1999; Escoffre et al. 2011). This is correlated to the fact that the anode-facing pole is electroporated first and has a lower threshold for the electric field strength required for electroporation (Gabriel and Teissié 1997, 1999).

It should be noted that the overall transfer efficiency is also affected by the viability of cells as the pulses that can be most efficient for the molecular delivery can also severely affect the survival of the cells (Wolf et al. 1994).

In the context of drug delivery, electroporation is already in clinical use as electrochemotherapy, a method of increased local anticancer drug delivery for bleomycin and cisplatin, anticancer drugs with high intrinsic toxicity but low plasma membrane permeability (Mir 1999). The first clinical trials for electrochemotherapy with bleomycin were performed in 1993 (Belehradek et al. 1993) and for electrochemotherapy with cisplatin in 1998 (Sersa et al. 1998). In 2006, standard operating procedures for electrochemotherapy were established (Marty et al. 2006).

In contrast to small molecule delivery, DNA electrotransfection is a complex multistep process. The DNA must be present in the electroporation medium during the application of electric pulses for the electrotransfection to occur (Eynard et al. 1992; Wolf et al. 1994; Klenchin et al. 2011). In an electric field, negatively charged DNA moves toward the anode due to electrophoresis. This leads to interaction of the DNA with the electroporated cell membrane on the cathode-facing side that is followed by formation of DNA-pore complexes (Golzio et al. 2002). It has been visually shown that the complex formation occurs in discrete membrane patches, presumably strongly electroporated areas, and not on the whole surface of the cell membrane (Golzio et al. 2002). The area of possible DNA-membrane interaction can be enlarged by applying several pulses or pulses with changing directions, which create changing cathode-facing side each time (Faurie et al. 2004). This method has been shown to increase the efficiency of DNA electrotransfer under both in vitro and in vivo conditions (Faurie et al. 2003). However, it should be noted that multidirectional impulses can only increase the electrotransfer efficiency if low enough frequency is used as it takes about a second for a stable DNA-membrane complex to form (Faurie et al. 2010). If pulses with opposite direction are applied before the stable complex is formed, DNA can be removed from the complex and reenter the solution, thus decreasing the amount of complexed DNA and the efficiency of DNA electrotransfer (Faurie et al. 2010). The strength of the interaction between DNA and the membrane is dependent on the concentration of magnesium ions – too low concentration will cause inefficient complex formation, while too high concentration will cause too strong interaction between DNA and the membrane, thus limiting the ability of DNA to cross the plasma membrane (Haberl et al. 2013). DNA crosses the plasma membrane 10–40 min after complex formation (Wu and Yuan 2011) through a mechanism which is not yet fully understood. Some of the hypotheses state that the DNA is directly translocated through the plasma membrane (Yu et al. 2012). Other data shows that clathrin- and caveolin- dependent endocytosis is at least partially responsible (Rosazza et al. 2013). The recent data suggests that several different methods of DNA transport into cell are utilized simultaneously, although the DNA mostly enters the cell via clathrin-mediated endocytosis (Wu and Yuan 2011; Chang et al. 2014; Pavlin and Kandušer 2015). The involvement of the cytoskeleton was also shown as disrupting actin cytoskeleton decreases the efficiency of DNA electrotransfer (Rosazza et al. 2011). Once inside, DNA is transported via microtubule network to the nucleus (Vaughan and Dean 2006; Rosazza et al. 2013, 2016).

Electric pulses that are sufficient for small molecule electrotransfer might not lead to efficient DNA electrotransfer (Wolf et al. 1994). Specifically, millisecond-length electric pulses are shown to cause the most efficient DNA electrotransfer (Rols and Teissié 1998). However, an increase in the pulse duration while the amplitude stays constant causes a decrease in cell viability (Canatella et al. 2001; Haberl et al. 2013). One of the ways to avoid this pitfall was introduced in the early 1990s (Sukharev et al. 1992) and later expanded in the early 2000s (Bureau et al. 2000; Šatkauskas et al. 2002, 2005). The method relies on the application of short, high-voltage pulses for permeabilization of plasma membrane, followed by millisecond-duration low-voltage pulse(s) to promote DNA electrophoresis and transfection efficiency. This protocol has provided good results under both in vivo (Bureau et al. 2000; Šatkauskas et al. 2002, 2005) and in vitro (Kanduser et al. 2009; Pavlin and Kandušer 2015) conditions. It should be noted that high enough plasmid concentrations under in vitro conditions can override the effect of low-voltage pulses (Kanduser et al. 2009; Čepurnienė et al. 2010). While the effect of addition of low-voltage pulse has been mainly attributed to the increased electrophoretic movement of highly negatively charged DNA (Šatkauskas et al. 2002, 2005), it has been also hypothesized that this effect is caused by the increased efficiency of DNA-membrane complex formation (Pavlin et al. 2010). It has been shown that weak electric fields can enhance macromolecule internalization, presumably via stimulation of endocytotic pathways (Antov et al. 2005).

The electrotransfection of plasmid DNA has a potential in clinical application. Till now, the research for clinical application on DNA electrotransfer has mainly focused on transferring immune system-activating molecules for the improvement of anticancer treatments (Calvet and Mir 2016). A clinical trial on electrotransfer of IL-12 coding plasmid has been launched in 2008 and reached moderate success, showing the potential of this application for cancer treatment (Daud et al. 2008).

5.9 Sonoporation

Drug and gene delivery to cells and tissues via sonoporation has received increasing attention during the past decade. The method of sonoporation employs the delivery of ultrasound (US) to cells and tissues in the presence of a US contrast agent – microbubbles (MB). The interaction of MBs with ultrasound results in MB cavitation that is thought to be the primary cause of the cell sonoporation (Fig. 5.1h). Two main types of cavitation, stable and inertial, produce a variety of biological effects.

At low acoustic pressures, stable cavitation, which is defined as periodic and linear MB oscillations around a constant bubble diameter, is the dominant mode of cavitation. Stable cavitation creates microstreaming – a circulating fluid flow around the bubble, which induces shear stress to the cell membrane which is proportional to the oscillation amplitude. When oscillation amplitudes are high, the associated shear forces are capable to directly rupture lipid vesicles (Marmottant and Hilgenfeldt 2003). Thus, it is suggested that shear forces induced by stable cavitation could be responsible for direct or indirect increase of the cell membrane permeability and molecular delivery (van Wamel et al. 2006; Kooiman et al. 2011; De Cock et al. 2015) by activating endocytotic processes or causing pore formation (Meijering et al. 2009; Juffermans et al. 2014; De Cock et al. 2015).

At higher acoustic pressures, the oscillation of MBs is accompanied by rapid growth of MB size, which ultimately results in a violent collapse of the bubble. Such MB behavior is called inertial cavitation. Main effects of inertial cavitation are microjets, shock waves, and drastic temperature increase in the surrounding medium (Didenko et al. 1999; Ohl et al. 2006; Kudo et al. 2009). MB inertial cavitation is described as the key mechanism to achieve an efficient cell permeabilization. This is due to stronger physical effects of bubble collapse, especially microjets (liquid jets travelling at the sonic speed), which can either induce catastrophic shear stress or directly pierce cell membranes (Ohl et al. 2006; Kudo et al. 2009).

As MBs violently collapse, they produce acoustic fields with components in broad frequency ranges. Thus, MB broadband noise is a direct evidence of MB inertial cavitation (Sundaram et al. 2003). It has been also demonstrated, at the single-cell level, that intracellular transmembrane current was directly associated with the onset of MB broadband noise (Zhou et al. 2008). The studies of implicit sonoporation dosimetry, designed to quantify inertial cavitation metrics, inertial cavitation dose, and MB sonodestruction rate, have shown molecular intracellular delivery (Lai et al. 2006; Qiu et al. 2010; Tamošiūnas et al. 2012b; Maciulevičius et al. 2015), cell viability (Lai et al. 2006; Qiu et al. 2010; Tamošiūnas et al. 2012b), and pore size (Qiu et al. 2010) to be directly associated with inertial cavitation according to strong correlation with metrics.

Ever since it has been shown that both stable and inertial cavitation is able to induce molecular sonotransfer, there has been an ongoing discussion as to which mode is more important for molecule delivery via sonoporation (Lentacker et al. 2014). The results obtained in the experiments at the single-cell level demonstrate the effects of stable cavitation to become more important when MBs appear in close vicinity to cells, as in case of MB attachment or targeting (Kooiman et al. 2011). Because increasing distances diminish the effects of microstreaming, e.g., in cell suspensions, inertial cavitation seems to be the dominant mechanism for efficient sonoporation when MBs are stochastically moving (Lai et al. 2006; Qiu et al. 2010; Tamošiūnas et al. 2012a, b; Maciulevičius et al. 2015).

There are two main hypotheses about the molecular transfer at the membrane level. The pore formation hypothesis suggests that molecules enter the cell interior through US-MB activity-induced membrane pores, while the endocytosis hypothesis declares that molecules enter the cell within membrane vesicles that liberate their cargo into the cytoplasm after degradation.

Various compounds of different types and sizes have been directly or indirectly observed to enter cells through pores: calcein (Ohl et al. 2006), propidium iodide (Kudo et al. 2009), dextrans (Mehier-Humbert et al. 2005a), nanospheres (Mehier-Humbert et al. 2005a), and DNA (Mehier-Humbert et al. 2005b; Qiu et al. 2010).

The direct evidence of pore existence is provided by microscopy studies, performed after sonoporation (Ohl et al. 2006; Qiu et al. 2010). Indirect evidence is obtained using other criteria, such as diffusive molecule distribution in cell cytoplasm after sonoporation (Mehier-Humbert et al. 2005a, b; Schlicher et al. 2006), fast DNA expression after sonoporation as opposed to lipofection (Mehier-Humbert et al. 2005b), and high molecular delivery rate achieved after endocytosis inhibition (Mehier-Humbert et al. 2005b; Meijering et al. 2009). A recent study (Fan et al. 2010) provides convincing results that indicated nonspecific sonopores, characterized by bidirectional, diffusion-driven transport of small molecules across the cell membrane.

Different authors report different putative pore sizes. A study of Zhou et al. (2009) has used voltage clamp technique to estimate a pore size of 110 ± 40 nm. By comparing sonotransfer efficiency of various size molecules, Mehier-Humbert et al. (2005a) report the maximal pore size to be around 75 nm. Using microscopy, ~1 μm pores were observed (Schlicher et al. 2006; Ohl et al. 2006; Kudo et al. 2009). Pore lifetime is estimated to vary from a few seconds to more than 1 min (Deng et al. 2004; Mehier-Humbert et al. 2005b; Schlicher et al. 2006). Cell membrane recovery occurs via exocytosis-like process, when membrane vesicles travel through the cytoplasm to damaged cell region and reform cell membrane (Schlicher et al. 2006). Calcium ions have been shown to accelerate or be crucial for membrane resealing (Kudo et al. 2009).

The alternative hypothesis is that molecules enter cells via endocytosis rather than pores. Evidence for this hypothesis comes from microscopy experiments by Duvshani-Eshet et al. (2006) and Zeghimi et al. (2015), who directly observed membrane changes after sonoporation, using atomic force (AFM) and scanning electron microscopy (SEM). According to these studies, the pit-like structural membrane changes resemble endosome forming endocytotic pits more closely than conductive pores.

The role of endocytosis has been investigated (Meijering et al. 2009; Juffermans et al. 2014) using dextrans and nucleic acids, respectively. It was stated that the role of endocytosis increases as the delivered molecules become larger (Meijering et al. 2009). Molecules greater than ~155 kDa enter cells mostly via endocytosis. Moreover, molecules greater than 500 kDa enter cells only via endocytosis. Small dextrans and siRNAs are homogeneously distributed in the cytoplasm, suggesting direct entry, while large dextrans and plasmids are distributed inhomogeneously, in a distinct-vehicle localization, suggesting endocytotic entrance. Also, much higher siRNA transfection and translation rate in comparison to plasmid DNA suggests no contact between siRNA and lysosomes. Inhibition of endocytosis either via ATP depletion, specific inhibition of clathrin- or caveolin-mediated endocytosis, or macropinocytosis decreases the amount of delivered dextrans of all molecular weights. This is especially apparent with 4.4 kDa dextran, and the delivery of 500 kDa dextran is effectively eliminated. There are a few hypotheses concerning the triggering mechanism of endocytosis. The reported increase in intracellular calcium ion concentration, reactive oxygen species, temperature increase, and shear forces are described as possible triggers (Lentacker et al. 2014).

Recent studies performed with anticancer drugs have shown sonoporation to significantly enhance the delivery of various anticancer drugs to cancer cells, thus showing the high potential of sonoporation as a tool for cancer treatment. Different studies have employed sonoporation to enhance the cytotoxicity of different drugs in vitro; the main drugs used in these studies were bleomycin (Tamošiūnas et al. 2012a, b, 2016; Lamanauskas et al. 2013) and doxorubicin (Lentacker et al. 2010; Geers et al. 2011; Maciulevičius et al. 2015). Drug sonotransfer research in vitro primarily aims to advance the anticancer drug sonotransfer efficiency to cells and to determine the optimal conditions for sonoporation in order to establish this technique as an effective tool for anticancer therapy in vivo.

Tamošiūnas et al. have reported reversible BLM and DOX delivery to Chinese hamster ovary cells to be ~30%, while cell viability was ~70% (Tamošiūnas et al. 2012a, b), and Maciulevičius et al. have reported DOX delivery of ~20% and cell viability of ~60% (Maciulevičius et al. 2015). During the experiments the dominant mechanism is reported to be inertial cavitation as indicated by direct broadband noise and MB concentration decay measurements.

Lentacker et al. and Geers et al. have developed doxorubicin-liposome-loaded-microbubbles, which, in combination with US, have diminished cancer cell viability in vitro down to ~20–10% (Lentacker et al. 2010; Geers et al. 2011). Moreover, the cytotoxic effect has been achieved with very low doxorubicin doses (5.4 μg/ml) (Geers et al. 2011). Later studies have tested the efficiency of MBs bound to paclitaxel-loaded liposomes (Yan et al. 2013). After US exposure, these agents have increased drug accumulation in breast cancer tumors, resulting in strong tumor inhibition and reduced cytotoxicity to the liver and kidney.

It has been shown that, upon an intravenous injection, nanodroplets escape vascularization due to enhanced permeability and retention effect and coalesce in the whole tumor parenchyma to produce MBs with a strong echocontrast (Rapoport et al. 2007). MBs preformed from nanodroplets loaded with DOX or paclitaxel have been shown to enhance the drug uptake into tumor cells in vitro and to significantly reduce tumor growth or strong tumor regression (~5 fold) in vivo, in ovarian and pancreatic tumors (Rapoport et al. 2007, 2009).

Sonoporation can become a useful tool to treat brain malignancies as focused US can be exploited to disrupt the blood-brain barrier for therapeutics delivery to gliomas (Ting et al. 2012; Treat et al. 2012).

The first clinical trial for in vivo drug delivery using sonoporation has been performed in 2013 (Kotopoulis et al. 2013). In this trial, SonoVue® MBs and the anticancer drug gemcitabine have been administered intravenously. All five treated patients have shown reduced tumor growth; in two of them, the maximum tumor diameter has been temporally decreased to 80 ± 5% and permanently to 70 ± 5% as compared to the original size. These results show that sonoporation can significantly enhance the effect of standard chemotherapy procedures. Additional studies from the same research group applied gemcitabine and sonoporation to ten patients with inoperable pancreatic cancer and reported that the maximum tumor diameter decreased in five patients with median patient survival increase from 8.9 months to 17.6 months (Dimcevski et al. 2016).

Sonoporation can also be used for gene therapy, i.e., delivery of exogenous genes to cells for the production of desired proteins. Mehier-Humbert et al. obtained 30–40% transfection efficiency, which increases with time after sonoporation, reaching up to ~55% within 2–5 h after sonoporation (Mehier-Humbert et al. 2005b). Juffermans et al. have obtained 43.0 ± 4.2% transfer with plasmid DNA and 97.9 ± 1.5% with siRNA, respectively (Juffermans et al. 2014). The authors associate the high uptake of siRNA with their direct entry into the cell cytoplasm through pores and the lower uptake of plasmid DNA with endocytotic transport through the plasma membrane. To our knowledge, the highest reported rate of cell transfection using sonoporation was 70%, while sustaining ~85% cell viability (Escoffre et al. 2013b). This result has been obtained using Vevo Micromarker MBs, which exhibited the highest rate of attenuation as compared to SonoVue® or BR14 MBs.

Sonoporation in vivo can be used for cancer and cardiovascular disease treatment. Anticancer effects are induced by oncogene inactivation, replacement of deficient tumor suppressor genes, or the introduction of genes that enhance cytotoxic effects and stimulate immune response. For cardiovascular disease treatment, angiogenesis-stimulating factor genes could be introduced (Escoffre et al. 2013a).

Another promising method for cancer treatment is gene suppression using siRNA. The main purpose of small siRNA molecule delivery to tumor tissue is the silencing of anti-apoptotic genes, thus causing cell apoptosis (Yin et al. 2013, 2014). Apoptotic cell fragments are phagocytized, reducing inflammatory response and diminishing harmful effects to surrounding cells.

The recent research in the field of gene therapy using sonoporation has shown efficient therapeutic results using naked DNA. Various studies reported a reduction of liver tumor growth by ~75% (Sakakima et al. 2005) and of prostate cancer growth by ~80%, after repeated treatment (Duvshani-Eshet et al. 2006). The survival of fibrosarcoma tumor-bearing mice has been prolonged for up to 160 days after treatment, and resistance for repeated tumor induction has developed in the treated animals (Casey et al. 2010). As shown in rats, the capillary density in the ischemic myocardium has increased twofold after vascular endothelial growth factor (VEGF) gene delivery (Zhigang et al. 2004).

As compared to chemotherapeutics, the introduction of genes in vivo poses lower cytotoxicity risk. However, gene therapy causes another problem, namely, degradation of DNA/RNA in the serum. Thus, the concentration of nucleic acids has to be increased in order to obtain efficient bioeffects. One way to overcome this obstacle is nucleic acid loading onto MBs while also applying MB targeting and reducing their size.

DNA loading onto cationic MBs has shown higher ultrasound-mediated gene delivery and lower level of DNA degradation by DNAses both in vitro and in vivo studies (Panje et al. 2012; Wang et al. 2012). Luciferase plasmid-bearing MBs, targeted to VCAM-1 and mucosal addressin cell adhesion molecule-1 (MAdCAM-1), have shown several (3.6 and 8.5)-fold higher MB accumulation in comparison to the nontargeted MB form in the ileocecal area in the murine model. Pronounced plasmid DNA transfection, localized to mouse gastrointestinal tract with no detectable luminescence in the heart, liver, spleen, or kidney, can be achieved after gene delivery with sonoporation (Tlaxca et al. 2013).

Suzuki et al. have synthesized liposomes (0.5–1 μm), encapsulating lipid nanobubbles, named "bubble liposomes" (Suzuki et al. 2010). The latter are prominently smaller than commercial microbubbles. A study showed the possibility to use US in tandem with bubble liposomes to deliver interleukin 12 (IL-12)-encoding plasmids to mice bearing murine ovarian carcinoma; after repetitive treatment, complete tumor regression has been achieved in 80% of cases (Suzuki et al. 2010). Endo-Takahashi et al. have designed bubble liposomes, loaded with microRNA or plasmid DNA (Endo-Takahashi et al. 2013, 2014). These novel agents have shown pronounced theranostic characteristics; they are capable to provide both the strong echocontrast and cargo protection as well as targeted delivery. The efficient nucleic acid delivery to mouse ischemic hind limb has led to enhanced angiogenesis as well as improved blood flow.

NBs, electrostatically coated with siRNA-loaded micelles, have been successfully applied in vivo to silence anti-apoptosis gene sirtuin-2 (SIRT2) and significantly inhibit glioma growth in mouse model (>2 fold compared control group) (Yin et al. 2013). Subsequent siRNA micelle and paclitaxel co-loading onto NBs has elicited strong tumor growth suppression to initial tumor level due to the synergistic effect of anti-apoptosis gene B-cell lymphoma-2 (BCL-2) silencing and drug cytotoxic activity (Yin et al. 2014).

5.10 Conclusion

There are many physical methods for drug and gene delivery. Each of these has their own advantages and drawbacks, which are summarized in Table 5.1. Microinjection can achieve very high transfection yield, but it is limited to single-cell treatment. Ballistic gene delivery allows rather efficient transfection but is limited to the surface and may cause damage to cells and tissues. Hydrodynamic gene delivery causes very efficient gene delivery to rodents but raises serious safety concerns, thus limiting the translation of the method to human clinical use. Electroporation is a safe, clinically approved method for drug delivery that also achieves reasonable DNA transfection efficiency; however, deep tissue electroporation is highly invasive.

Table 5.1 Features of different physical cell permeabilization methods

Delivery technique	System	Drug delivery efficiency	Gene delivery efficiency	Cell survival	Safety	Tissue penetration	Invasiveness
Ballistic delivery	In vitro, in vivo	N/A	Low	Medium to high	Medium to high	Low	Low
Electroporation	In vitro, in vivo	High	Low to medium	Medium to high	High	Medium	High
Electrospray	In vitro, in vivo	High	Low to medium	Medium	High	Low	Low to medium
Hydrodynamic delivery	In vivo	N/A	Medium	N/A	Low	Medium	High
Magnetoporation	In vitro, in vivo	Medium	Low	Low to medium	High	Medium	Low
Microinjection	Single cells only	N/A	High	High	N/A	N/A	N/A
Photoporation	In vitro, in vivo	Medium to high	Medium to high	Medium	Medium	Low to medium	Medium
Sonoporation	In vitro, in vivo	Medium	Low	Medium	Medium	High	Low

Here *N/A* not applicable. Safety of method refers to short- and long-term damage to whole organism

Sonoporation can be employed to combine ultrasound diagnostics with drug delivery, but molecular delivery rates are lower than desired for clinical use. Other emerging techniques, such as magnetoporation, photoporation, and electrospray, have to be developed further before they can be applied in clinical trials for drug and gene delivery.

Acknowledgments This work was supported by the grants SVE-08, 2014 and MIP-034/2013 from the Research Council of Lithuania. The authors would also like to acknowledge networking support by the COST Action TD1104.

References

Antov Y, Barbul A, Mantsur H, Korenstein R (2005) Electroendocytosis: exposure of cells to pulsed low electric fields enhances adsorption and uptake of macromolecules. Biophys J 88:2206–2223. doi:10.1529/biophysj.104.051268

Balmayor ER, Azevedo HS, Reis RL (2011) Controlled delivery systems: from pharmaceuticals to cells and genes. Pharm Res 28:1241–1258. doi:10.1007/s11095-011-0392-y

Belehradek M, Domenge C, Luboinski B et al (1993) Electrochemotherapy, a new antitumor treatment. First clinical phase I-II trial. Cancer 72:3694–3700

Bernhardt J, Pauly H (1973) On the generation of potential differences across the membranes of ellipsoidal cells in an alternating electrical field. Biophysik 10:89–98

Bonamassa B, Hai L, Liu D (2011) Hydrodynamic gene delivery and its applications in pharmaceutical research. Pharm Res 28:694–701. doi:10.1007/s11095-010-0338-9

Boulaiz H, Marchal JA, Prados J et al (2005) Non-viral and viral vectors for gene therapy. Cell Mol Biol (Noisy-le-Grand) 51:3–22

Bureau MF, Gehl J, Deleuze V et al (2000) Importance of association between permeabilization and electrophoretic forces for intramuscular DNA electrotransfer. Biochim Biophys Acta 1474:353–359

Calvet CY, Mir LM (2016) The promising alliance of anti-cancer electrochemotherapy with immunotherapy. Cancer Metastasis Rev 35:165–177. doi:10.1007/s10555-016-9615-3

Canatella PJ, Karr JF, Petros JA, Prausnitz MR (2001) Quantitative study of electroporation-mediated molecular uptake and cell viability. Biophys J 80:755–764. doi:10.1016/S0006-3495(01)76055-9

Casey G, Cashman JP, Morrissey D et al (2010) Sonoporation mediated immunogene therapy of solid tumors. Ultrasound Med Biol 36:430–440. doi:10.1016/j.ultrasmedbio.2009.11.005

Celis JE (1977) Injection of tRNAs into somatic cells: search for in vivo systems to assay potential nonsense mutations in somatic cells. Brookhaven Symp Biol 29:178–196

Čepurnienė K, Ruzgys P, Treinys R et al (2010) Influence of plasmid concentration on DNA electrotransfer in vitro using high-voltage and low-voltage pulses. J Membr Biol 236:81–85. doi:10.1007/s00232-010-9270-5

Chang C-C, Wu M, Yuan F (2014) Role of specific endocytic pathways in electrotransfection of cells. Mol Ther Methods Clin Dev 1:14058. doi:10.1038/mtm.2014.58

Chen D-R, Wendt CH, Pui DYH (2000) A novel approach for introducing bio-materials into cells. J Nanopart Res 2:133–139. doi:10.1023/A:1010084032006

Chen J-Y, Liao Y-L, Wang T-H, Lee W-C (2006) Transformation of Escherichia coli mediated by magnetic nanoparticles in pulsed magnetic field. Enzym Microb Technol 39:366–370. doi:10.1016/j.enzmictec.2005.11.035

Das AK, Gupta P, Chakraborty D (2015) Physical methods of gene transfer: Kinetics of gene delivery into cells: A Review. Agric Rev 36:61. doi:10.5958/0976-0741.2015.00007.0

Daud AI, DeConti RC, Andrews S et al (2008) Phase I trial of interleukin-12 plasmid electroporation in patients with metastatic melanoma. J Clin Oncol 26:5896–5903. doi:10.1200/JCO.2007.15.6794

De Cock I, Zagato E, Braeckmans K et al (2015) Ultrasound and microbubble mediated drug delivery: acoustic pressure as determinant for uptake via membrane pores or endocytosis. J Control Release 197:20–28. doi:10.1016/j.jconrel.2014.10.031
Deng CX, Sieling F, Pan H, Cui J (2004) Ultrasound-induced cell membrane porosity. Ultrasound Med Biol 30:519–526. doi:10.1016/j.ultrasmedbio.2004.01.005
Didenko YT, McNamara III WB, Suslick KS (1999) Hot spot conditions during cavitation in water. J Am Chem Soc. doi: 10.1021/JA9844635
Dimcevski G, Kotopoulis S, Bjånes T et al (2016) A human clinical trial using ultrasound and microbubbles to enhance gemcitabine treatment of inoperable pancreatic cancer. J Control Release 243:172–181. doi:10.1016/j.jconrel.2016.10.007
Djuzenova CS, Zimmermann U, Frank H et al (1996) Effect of medium conductivity and composition on the uptake of propidium iodide into electropermeabilized myeloma cells. Biochim Biophys Acta 1284:143–152
Doukas AG, Flotte TJ, Akhamanov S et al (1996) Physical characteristics and biological effects of laser-induced stress waves. Ultrasound Med Biol 22:151–164. doi:10.1016/0301-5629(95)02026-8
Duvshani-Eshet M, Baruch L, Kesselman E et al (2006) Therapeutic ultrasound-mediated DNA to cell and nucleus: bioeffects revealed by confocal and atomic force microscopy. Gene Ther 13:163–172. doi:10.1038/sj.gt.3302642
Endo-Takahashi Y, Negishi Y, Nakamura A et al (2013) pDNA-loaded bubble liposomes as potential ultrasound imaging and gene delivery agents. Biomaterials 34:2807–2813. doi:10.1016/j.biomaterials.2012.12.018
Endo-Takahashi Y, Negishi Y, Nakamura A et al (2014) Systemic delivery of miR-126 by miRNA-loaded bubble liposomes for the treatment of hindlimb ischemia. Sci Rep 4:3883. doi:10.1038/srep03883
Escoffre J-M, Portet T, Favard C et al (2011) Electromediated formation of DNA complexes with cell membranes and its consequences for gene delivery. Biochim Biophys Acta Biomembr 1808:1538–1543. doi:10.1016/j.bbamem.2010.10.009
Escoffre J-M, Novell A, Piron J et al (2013a) Microbubble attenuation and destruction: are they involved in sonoporation efficiency? IEEE Trans Ultrason Ferroelectr Freq Control 60:46–52. doi:10.1109/TUFFC.2013.2536
Escoffre J-M, Zeghimi A, Novell A, Bouakaz A (2013b) In-vivo gene delivery by sonoporation: recent progress and prospects. Curr Gene Ther 13:2–14
Eynard N, Sixou S, Duran N, Teissie J (1992) Fast kinetics studies of Escherichia coli electrotransformation. Eur J Biochem 209:431–436
Fan Z, Kumon RE, Park J, Deng CX (2010) Intracellular delivery and calcium transients generated in sonoporation facilitated by microbubbles. J Control Release 142:31–39. doi:10.1016/j.jconrel.2009.09.031
Faurie C, Golzio M, Moller P et al (2003) Cell and animal imaging of electrically mediated gene transfer. DNA Cell Biol 22:777–783. doi:10.1089/104454903322624984
Faurie C, Phez E, Golzio M et al (2004) Effect of electric field vectoriality on electrically mediated gene delivery in mammalian cells. Biochim Biophys Acta Biomembr 1665:92–100. doi:10.1016/j.bbamem.2004.06.018
Faurie C, Rebersek M, Golzio M et al (2010) Electro-mediated gene transfer and expression are controlled by the life-time of DNA/membrane complex formation. J Gene Med 12:117–125. doi:10.1002/jgm.1414
Fraites T, Mah CS, Zolotukhin I (2000) Microsphere-mediated delivery of recombinant AAV vectors in vitro and in vivo. Mol Ther 1:S239
Gabriel B, Teissié J (1997) Direct observation in the millisecond time range of fluorescent molecule asymmetrical interaction with the electropermeabilized cell membrane. Biophys J 73:2630–2637. doi:10.1016/S0006-3495(97)78292-4
Gabriel B, Teissié J (1998) Mammalian cell electropermeabilization as revealed by millisecond imaging of fluorescence changes of ethidium bromide in interaction with the membrane. Bioelectrochem Bioenerg 47:113–118. doi:10.1016/S0302-4598(98)00174-3
Gabriel B, Teissié J (1999) Time courses of mammalian cell electropermeabilization observed by millisecond imaging of membrane property changes during the pulse. Biophys J 76:2158–2165. doi:10.1016/S0006-3495(99)77370-4

Geers B, Lentacker I, Sanders NN et al (2011) Self-assembled liposome-loaded microbubbles: the missing link for safe and efficient ultrasound triggered drug-delivery. J Control Release 152:249–256. doi:10.1016/j.jconrel.2011.02.024

Golzio M, Teissié J, Rols M-P (2002) Cell synchronization effect on mammalian cell permeabilization and gene delivery by electric field. Biochim Biophys Acta 1563:23–28

Haberl S, Kandušer M, Flisar K et al (2013) Effect of different parameters used for in vitro gene electrotransfer on gene expression efficiency, cell viability and visualization of plasmid DNA at the membrane level. J Gene Med 15:169–181. doi:10.1002/jgm.2706

Hibino M, Shigemori M, Itoh H et al (1991) Membrane conductance of an electroporated cell analyzed by submicrosecond imaging of transmembrane potential. Biophys J 59:209–220. doi:10.1016/S0006-3495(91)82212-3

Hibino M, Itoh H, Kinosita K (1993) Time courses of cell electroporation as revealed by submicrosecond imaging of transmembrane potential. Biophys J 64:1789–1800. doi:10.1016/S0006-3495(93)81550-9

Hooper JW, Kamrud KI, Elgh F et al (1999) DNA vaccination with hantavirus M segment elicits neutralizing antibodies and protects against Seoul virus infection. Virology 255:269–278. doi:10.1006/viro.1998.9586

Hradetzky D, Boehringer S, Geiser T, Gazdhar A (2012) An approach towards bronchoscopic-based gene therapy using electrical field accelerated plasmid droplets. In: 2012 Annual International Conference of the IEEE Engineering in Medicine and Biology Society. IEEE, pp 5753–5756

Ikemoto K, Sakata I, Sakai T (2012) Collision of millimetre droplets induces DNA and protein transfection into cells. Sci Rep. doi:10.1038/srep00289

Irvine KR, Rao JB, Rosenberg SA, Restifo NP (1996) Cytokine enhancement of DNA immunization leads to effective treatment of established pulmonary metastases. J Immunol 156:238–245

Juffermans LJM, Meijering BDM, Henning RH, Deelman LE (2014) Ultrasound and microbubble-targeted delivery of small interfering RNA into primary endothelial cells is more effective than delivery of plasmid DNA. Ultrasound Med Biol 40:532–540. doi:10.1016/j.ultrasmedbio.2013.09.031

Kanduser M, Miklavcic D, Pavlin M (2009) Mechanisms involved in gene electrotransfer using high- and low-voltage pulses – an in vitro study. Bioelectrochemistry 74:265–271. doi:10.1016/j.bioelechem.2008.09.002

Kardos TJ, Rabussay DP (2012) Contactless magneto-permeabilization for intracellular plasmid DNA delivery in-vivo. Hum Vaccin Immunother 8:1707–1713. doi:10.4161/hv.21576

Klatzmann D, Chérin P, Bensimon G et al (1998) A phase I/II dose-escalation study of herpes simplex virus type 1 thymidine kinase "suicide" gene therapy for metastatic melanoma. Hum Gene Ther 9:2585–2594. doi:10.1089/hum.1998.9.17-2585

Klein TM, Wolf ED, Wu R, Sanford JC (1987) High-velocity microprojectiles for delivering nucleic acids into living cells. Nature 327:70–73. doi:10.1038/327070a0

Klenchin VA, Frye JJ, Jones MH et al (2011) Structure-function analysis of the C-terminal domain of CNM67, a core component of the Saccharomyces cerevisiae spindle pole body. J Biol Chem 286:18240–18250. doi:10.1074/jbc.M111.227371

Kobayashi N, Nishikawa M, Hirata K, Takakura Y (2004) Hydrodynamics-based procedure involves transient hyperpermeability in the hepatic cellular membrane: implication of a nonspecific process in efficient intracellular gene delivery. J Gene Med 6:584–592. doi:10.1002/jgm.541

Kooiman K, Foppen-Harteveld M, van der Steen AFW, de Jong N (2011) Sonoporation of endothelial cells by vibrating targeted microbubbles. J Control Release 154:35–41. doi:10.1016/j.jconrel.2011.04.008

Kotopoulis S, Dimcevski G, Gilja OH et al (2013) Treatment of human pancreatic cancer using combined ultrasound, microbubbles, and gemcitabine: a clinical case study. Med Phys 40:72902. doi:10.1118/1.4808149

Kudo N, Okada K, Yamamoto K (2009) Sonoporation by single-shot pulsed ultrasound with microbubbles adjacent to cells. Biophys J 96:4866–4876. doi:10.1016/j.bpj.2009.02.072

Lai C-Y, Wu C-H, Chen C-C, Li P-C (2006) Quantitative relations of acoustic inertial cavitation with sonoporation and cell viability. Ultrasound Med Biol 32:1931–1941. doi:10.1016/j.ultrasmedbio.2006.06.020

Lakshmanan S, Gupta GK, Avci P et al (2014) Physical energy for drug delivery; poration, concentration and activation. Adv Drug Deliv Rev 71:98–114. doi:10.1016/j.addr.2013.05.010

Lamanauskas N, Novell A, Escoffre J-M et al (2013) Bleomycin delivery into cancer cells *in vitro* with ultrasound and SonoVue® or BR14® microbubbles. J Drug Target 21:407–414. doi:10.3109/1061186X.2012.761223

Lentacker I, Geers B, Demeester J et al (2010) Design and evaluation of doxorubicin-containing microbubbles for ultrasound-triggered doxorubicin delivery: cytotoxicity and mechanisms involved. Mol Ther 18:101–108. doi:10.1038/mt.2009.160

Lentacker I, De Cock I, Deckers R et al (2014) Understanding ultrasound induced sonoporation: definitions and underlying mechanisms. Adv Drug Deliv Rev 72:49–64. doi:10.1016/j.addr.2013.11.008

Lin MTS, Pulkkinen L, Uitto J, Yoon K (2000) The gene gun: current applications in cutaneous gene therapy. Int J Dermatol 39:161–170. doi:10.1046/j.1365-4362.2000.00925.x

Liu F, Song YK, Liu D (1999) Hydrodynamics-based transfection in animals by systemic administration of plasmid DNA. Gene Ther 6:1258–1266. doi:10.1038/sj.gt.3300947

Liu D, Wang L, Wang Z, Cuschieri A (2012) Magnetoporation and magnetolysis of cancer cells via carbon nanotubes induced by rotating magnetic fields. Nano Lett 12:5117–5121. doi:10.1021/nl301928z

Maciulevičius M, Tamošiūnas M, Jurkonis R et al (2015) Analysis of metrics for molecular Sonotransfer in vitro. Mol Pharm 12:3620–3627. doi:10.1021/acs.molpharmaceut.5b00347

Markovskaja S, Novickij V, Grainys A, Novickij J (2014) Irreversible magnetoporation of micro-organisms in high pulsed magnetic fields. IET Nanobiotechnol 8:157–162. doi:10.1049/iet-nbt.2013.0005

Marmottant P, Hilgenfeldt S (2003) Controlled vesicle deformation and lysis by single oscillating bubbles. Nature 423:153–156. doi:10.1038/nature01613

Marty M, Sersa G, Garbay JR et al (2006) Electrochemotherapy – an easy, highly effective and safe treatment of cutaneous and subcutaneous metastases: results of ESOPE (European standard operating procedures of Electrochemotherapy) study. Eur J Cancer Suppl 4:3–13. doi:10.1016/j.ejcsup.2006.08.002

Mehier-Humbert S, Bettinger T, Yan F, Guy RH (2005a) Plasma membrane poration induced by ultrasound exposure: implication for drug delivery. J Control Release 104:213–222. doi:10.1016/j.jconrel.2005.01.007

Mehier-Humbert S, Bettinger T, Yan F, Guy RH (2005b) Ultrasound-mediated gene delivery: kinetics of plasmid internalization and gene expression. J Control Release 104:203–211. doi:10.1016/j.jconrel.2005.01.011

Meijering BDM, Juffermans LJM, van Wamel A et al (2009) Ultrasound and microbubble-targeted delivery of macromolecules is regulated by induction of endocytosis and pore formation. Circ Res 104:679–687. doi:10.1161/CIRCRESAHA.108.183806

Mir O (1999) Mechanisms of electrochemotherapy. Adv Drug Deliv Rev 35:107–118

Mohanty SK, Sharma M, Gupta PK (2003) Laser-assisted microinjection into targeted animal cells. Biotechnol Lett 25:895–899

Namgung R, Singha K, Yu MK et al (2010) Hybrid superparamagnetic iron oxide nanoparticle-branched polyethylenimine magnetoplexes for gene transfection of vascular endothelial cells. Biomaterials 31:4204–4213. doi:10.1016/j.biomaterials.2010.01.123

Neumann E, Rosenheck K (1972) Permeability changes induced by electric impulses in vesicular membranes. J Membr Biol 10:279–290

Noack J, Vogel A (1995) Streak-photographic investigation of shock wave emission after laser-induced plasma formation in water. In: Jacques SL (ed). International Society for Optics and Photonics, p 284

Ohl C-D, Arora M, Ikink R et al (2006) Sonoporation from jetting cavitation bubbles. Biophys J 91:4285–4295. doi:10.1529/biophysj.105.075366

Okubo Y, Ikemoto K, Koike K et al (2008) DNA introduction into living cells by water droplet impact with an electrospray process. Angew Chemie Int Ed 47:1429–1431. doi:10.1002/anie.200704429

Panje CM, Wang DS, Pysz MA et al (2012) Ultrasound-mediated gene delivery with cationic versus neutral microbubbles: effect of DNA and microbubble dose on in vivo transfection efficiency. Theranostics 2:1078–1091. doi:10.7150/thno.4240

Paterson L, Agate B, Comrie M et al (2005) Photoporation and cell transfection using a violet diode laser. Opt Express 13:595–600

Pavlin M, Kandušer M (2015) New insights into the mechanisms of gene electrotransfer – experimental and theoretical analysis. Sci Rep 5:9132. doi:10.1038/srep09132

Pavlin M, Flisar K, Kandušer M (2010) The role of electrophoresis in gene electrotransfer. J Membr Biol 236:75–79. doi:10.1007/s00232-010-9276-z

Pitsillides CM, Joe EK, Wei X et al (2003) Selective cell targeting with light-absorbing microparticles and nanoparticles. Biophys J 84:4023–4032. doi:10.1016/S0006-3495(03)75128-5

Pucihar G, Kotnik T, Miklavcic D, Teissié J (2008) Kinetics of transmembrane transport of small molecules into electropermeabilized cells. Biophys J 95:2837–2848. doi:10.1529/biophysj.108.135541

Qiu Y, Luo Y, Zhang Y et al (2010) The correlation between acoustic cavitation and sonoporation involved in ultrasound-mediated DNA transfection with polyethylenimine (PEI) in vitro. J Control Release 145:40–48. doi:10.1016/j.jconrel.2010.04.010

Rapoport N, Gao Z, Kennedy A (2007) Multifunctional nanoparticles for combining ultrasonic tumor imaging and targeted chemotherapy. JNCI J Natl Cancer Inst 99:1095–1106. doi:10.1093/jnci/djm043

Rapoport NY, Kennedy AM, Shea JE et al (2009) Controlled and targeted tumor chemotherapy by ultrasound-activated nanoemulsions/microbubbles. J Control Release 138:268–276. doi:10.1016/j.jconrel.2009.05.026

Rols MP, Teissié J (1990) Electropermeabilization of mammalian cells. Quantitative analysis of the phenomenon. Biophys J 58:1089–1098. doi:10.1016/S0006-3495(90)82451-6

Rols M-P, Teissié J (1998) Electropermeabilization of mammalian cells to macromolecules: control by pulse duration. Biophys J 75:1415–1423. doi:10.1016/S0006-3495(98)74060-3

Rosazza C, Escoffre J-M, Zumbusch A, Rols M-P (2011) The actin cytoskeleton has an active role in the electrotransfer of plasmid DNA in mammalian cells. Mol Ther 19:913–921. doi:10.1038/mt.2010.303

Rosazza C, Buntz A, Rieß T et al (2013) Intracellular tracking of single-plasmid DNA particles after delivery by electroporation. Mol Ther 21:2217–2226. doi:10.1038/mt.2013.182

Rosazza C, Deschout H, Buntz A et al (2016) Endocytosis and endosomal trafficking of DNA after gene electrotransfer in vitro. Mol Ther Nucleic Acids 5:e286. doi:10.1038/mtna.2015.59

Sadik MM, Li J, Shan JW et al (2013) Quantification of propidium iodide delivery using millisecond electric pulses: experiments. Biochim Biophys Acta 1828:1322–1328. doi:10.1016/j.bbamem.2013.01.002

Sakakima Y, Hayashi S, Yagi Y et al (2005) Gene therapy for hepatocellular carcinoma using sonoporation enhanced by contrast agents. Cancer Gene Ther 12:884–889. doi:10.1038/sj.cgt.7700850

Šatkauskas S, Bureau MF, Puc M et al (2002) Mechanisms of in vivo DNA electrotransfer: respective contributions of cell electropermeabilization and DNA electrophoresis. Mol Ther 5:133–140. doi:10.1006/mthe.2002.0526

Šatkauskas S, André F, Bureau MF et al (2005) Electrophoretic component of electric pulses determines the efficacy of *in vivo* DNA electrotransfer. Hum Gene Ther 16:1194–1201. doi:10.1089/hum.2005.16.1194

Schlicher RK, Radhakrishna H, Tolentino TP et al (2006) Mechanism of intracellular delivery by acoustic cavitation. Ultrasound Med Biol 32:915–924. doi:10.1016/j.ultrasmedbio.2006.02.1416

Sersa G, Stabuc B, Cemazar M et al (1998) Electrochemotherapy with cisplatin: potentiation of local cisplatin antitumour effectiveness by application of electric pulses in cancer patients. Eur J Cancer 34:1213–1218

Suda T, Liu D (2007) Hydrodynamic gene delivery: its principles and applications. Mol Ther 15:2063–2069. doi:10.1038/sj.mt.6300314

Sukharev SI, Klenchin VA, Serov SM et al (1992) Electroporation and electrophoretic DNA transfer into cells. The effect of DNA interaction with electropores. Biophys J 63:1320–1327. doi:10.1016/S0006-3495(92)81709-5

Sundaram J, Mellein BR, Mitragotri S (2003) An experimental and theoretical analysis of ultrasound-induced permeabilization of cell membranes. Biophys J 84:3087–3101. doi:10.1016/S0006-3495(03)70034-4

Suzuki R, Namai E, Oda Y et al (2010) Cancer gene therapy by IL-12 gene delivery using liposomal bubbles and tumoral ultrasound exposure. J Control Release 142:245–250. doi:10.1016/j.jconrel.2009.10.027

Tamošiūnas M, Jurkonis R, Mir LM et al (2012a) Adjustment of ultrasound exposure duration to microbubble sonodestruction kinetics for optimal cell sonoporation in vitro. Technol Cancer Res Treat 11:375–387. doi:10.7785/tcrt.2012.500285

Tamošiūnas M, Jurkonis R, Mir LM et al (2012b) Microbubble sonodestruction rate as a metric to evaluate sonoporation efficiency. J Ultrasound Med 31:1993–2000

Tamošiūnas M, Mir LM, Chen W-S et al (2016) Intracellular delivery of bleomycin by combined application of electroporation and sonoporation in vitro. J Membr Biol 249:677–689. doi:10.1007/s00232-016-9911-4

Tao W, Wilkinson J, Stanbridge EJ, Berns MW (1987) Direct gene transfer into human cultured cells facilitated by laser micropuncture of the cell membrane. Proc Natl Acad Sci U S A 84:4180–4184

Teissié J, Rols MP (1993) An experimental evaluation of the critical potential difference inducing cell membrane electropermeabilization. Biophys J 65:409–413. doi:10.1016/S0006-3495(93)81052-X

Tekle E, Astumian RD, Chock PB (1994) Selective and asymmetric molecular transport across electroporated cell membranes. Proc Natl Acad Sci U S A 91:11512–11516

Ting C-Y, Fan C-H, Liu H-L et al (2012) Concurrent blood-brain barrier opening and local drug delivery using drug-carrying microbubbles and focused ultrasound for brain glioma treatment. Biomaterials 33:704–712. doi:10.1016/j.biomaterials.2011.09.096

Tiwari G, Tiwari R, Sriwastawa B et al (2012) Drug delivery systems: an updated review. Int J Pharm Investig 2:2–11. doi:10.4103/2230-973X.96920

Tlaxca JL, Rychak JJ, Ernst PB et al (2013) Ultrasound-based molecular imaging and specific gene delivery to mesenteric vasculature by endothelial adhesion molecule targeted microbubbles in a mouse model of Crohn's disease. J Control Release 165:216–225. doi:10.1016/j.jconrel.2012.10.021

Treat LH, McDannold N, Zhang Y et al (2012) Improved anti-tumor effect of liposomal doxorubicin after targeted blood-brain barrier disruption by MRI-guided focused ultrasound in rat glioma. Ultrasound Med Biol 38:1716–1725. doi:10.1016/j.ultrasmedbio.2012.04.015

Tresilwised N, Pithayanukul P, Mykhaylyk O et al (2010) Boosting oncolytic adenovirus potency with magnetic nanoparticles and magnetic force. Mol Pharm 7:1069–1089. doi:10.1021/mp100123t

Tsukakoshi M, Kurata S, Nomiya Y et al (1984) A novel method of DNA transfection by laser microbeam cell surgery. Appl Phys B Photophysics Laser Chem 35:135–140. doi:10.1007/BF00697702

Umebayashi Y, Miyamoto Y, Wakita M et al (2003) Elevation of plasma membrane permeability on laser irradiation of extracellular latex particles. J Biochem 134:219–224

Vaughan EE, Dean DA (2006) Intracellular trafficking of plasmids during transfection is mediated by microtubules. Mol Ther 13:422–428. doi:10.1016/j.ymthe.2005.10.004

Venslauskas MS, Šatkauskas S, Rodaitė-Riševičienė R (2010) Efficiency of the delivery of small charged molecules into cells in vitro. Bioelectrochemistry 79:130–135. doi:10.1016/j.bioelechem.2009.10.003

Vogel A, Busch S (1994) Time-resolved measurements of shock-wave emission and cavitation-bubble generation in intraocular laser surgery with ps- and ns-pulses. Springer Netherlands, New York, pp 105–117

van Wamel A, Kooiman K, Harteveld M et al (2006) Vibrating microbubbles poking individual cells: drug transfer into cells via sonoporation. J Control Release 112:149–155. doi:10.1016/j.jconrel.2006.02.007
Wang DS, Panje C, Pysz MA et al (2012) Cationic versus neutral microbubbles for ultrasound-mediated gene delivery in cancer. Radiology 264:721–732. doi:10.1148/radiol.12112368
Wang W, Li W, Ma N, Steinhoff G (2013) Non-viral gene delivery methods. Curr Pharm Biotechnol 14:46–60
Weaver JC, Chizmadzhev YA (1996) Theory of electroporation: a review. Bioelectrochem Bioenerg 41:135–160. doi:10.1016/S0302-4598(96)05062-3
Williams RS, Johnston SA, Riedy M et al (1991) Introduction of foreign genes into tissues of living mice by DNA-coated microprojectiles. Proc Natl Acad Sci U S A 88:2726–2730. doi:10.1073/PNAS.88.7.2726
Wolf H, Rols MP, Boldt E et al (1994) Control by pulse parameters of electric field-mediated gene transfer in mammalian cells. Biophys J 66:524–531
Wu M, Yuan F (2011) Membrane binding of plasmid DNA and endocytic pathways are involved in electrotransfection of mammalian cells. PLoS One 6:e20923. doi:10.1371/journal.pone.0020923
Yan F, Li L, Deng Z et al (2013) Paclitaxel-liposome-microbubble complexes as ultrasound-triggered therapeutic drug delivery carriers. J Control Release 166:246–255. doi:10.1016/j.jconrel.2012.12.025
Yao C-P, Zhang Z-X, Rahmanzadeh R, Huettmann G (2008) Laser-based gene transfection and gene therapy. IEEE Trans Nanobioscience 7:111–119. doi:10.1109/TNB.2008.2000742
Yin T, Wang P, Li J et al (2013) Ultrasound-sensitive siRNA-loaded nanobubbles formed by hetero-assembly of polymeric micelles and liposomes and their therapeutic effect in gliomas. Biomaterials 34:4532–4543. doi:10.1016/j.biomaterials.2013.02.067
Yin T, Wang P, Li J et al (2014) Tumor-penetrating codelivery of siRNA and paclitaxel with ultrasound-responsive nanobubbles hetero-assembled from polymeric micelles and liposomes. Biomaterials 35:5932–5943. doi:10.1016/j.biomaterials.2014.03.072
Yu M, Tan W, Lin H (2012) A stochastic model for DNA translocation through an electropore. Biochim Biophys Acta 1818:2494–2501. doi:10.1016/j.bbamem.2012.05.025
Zeghimi A, Escoffre JM, Bouakaz A (2015) Role of endocytosis in sonoporation-mediated membrane permeabilization and uptake of small molecules: a electron microscopy study. Phys Biol 12:66007. doi:10.1088/1478-3975/12/6/066007
Zeira E, Manevitch A, Khatchatouriants A et al (2003) Femtosecond infrared laser-an efficient and safe in vivo gene delivery system for prolonged expression. Mol Ther 8:342–350
Zhang G, Gao X, Song YK et al (2004) Hydroporation as the mechanism of hydrodynamic delivery. Gene Ther 11:675–682. doi:10.1038/sj.gt.3302210
Zhigang W, Zhiyu L, Haitao R et al (2004) Ultrasound-mediated microbubble destruction enhances VEGF gene delivery to the infarcted myocardium in rats. Clin Imaging 28:395–398. doi:10.1016/j.clinimag.2004.04.003
Zhou Y, Cui J, Deng CX (2008) Dynamics of sonoporation correlated with acoustic cavitation activities. Biophys J 94(7):L51–L53. doi: 10.1529/biophysj.107.125617
Zhou Y, Kumon RE, Cui J, Deng CX (2009) The size of sonoporation pores on the cell membrane. Ultrasound Med Biol 35:1756–1760. doi:10.1016/j.ultrasmedbio.2009.05.012

Chapter 6
Estrogen Receptors in Cell Membranes: Regulation and Signaling

Jolanta Saczko, Olga Michel, Agnieszka Chwiłkowska, Ewa Sawicka, Justyna Mączyńska, and Julita Kulbacka

Abstract Estrogens can stimulate the development, proliferation, migration, and survival of target cells. These biological effects are mediated through their action upon the plasma membrane estrogen receptors (ERs). ERs regulate transcriptional processes by nuclear translocation and binding to specific response elements, which leads to the regulation of gene expression. This effect is termed genomic or nuclear. However, estrogens may exert their biological activity also without direct binding to DNA and independently of gene transcription or protein synthesis. This action is called non-genomic or non-nuclear. Through non-genomic mechanisms, estrogens can modify regulatory cascades such as MAPK, P13K, and tyrosine cascade as well as membrane-associated molecules such as ion channels and G-protein-coupled receptors. The recent studies on the mechanisms of estrogen action provide an evidence that non-genomic and genomic effects converge. An example of such convergence is the potential possibility to modulate gene expression through these two independent pathways. The understanding of the plasma membrane estrogen receptors is crucial for the development of novel drugs and therapeutic protocols targeting specific receptor actions.

J. Saczko (✉) • O. Michel • A. Chwiłkowska • J. Mączyńska • J. Kulbacka
Department of Medical Biochemistry, Wroclaw Medical University,
Chalubinskiego 10, 50-368 Wroclaw, Poland
e-mail: jolanta.saczko@umed.wroc.pl

E. Sawicka
Department of Toxicology, Wroclaw Medical University,
Borowska 211, 50-552 Wroclaw, Poland

J. Kulbacka, S. Satkauskas (eds.), *Transport Across Natural and Modified Biological Membranes and its Implications in Physiology and Therapy*, Advances in Anatomy, Embryology and Cell Biology 227,
DOI 10.1007/978-3-319-56895-9_6

6.1 Introduction

Estrogens belong to steroid female sex hormones, which are the cholesterol derivatives. Estradiol (E2, 17β estradiol) shows the highest biological activity as compared to estrone (E1) and estriol (E3). Estrogens are synthesized in the ovary, placenta, testes, and adrenal cortex and are responsible for induction and development of secondary sexual features. Together with progesterone and gonadotropin hormones, they are involved in the regulation of the sexual cycle. In addition, they elicit multiple effects in other organs and tissues. Estrogens have a favorable impact on the lipid profile, increase the concentration of coagulation factors, and decrease the levels of insulin and glucose in blood. They are involved in stimulation of nitric oxide synthesis and in reduction of endothelin. They can modulate the immunological response and increase the membrane permeability. Moreover, estrogens play important roles in the development, proliferation, migration, and survival of target cells (Kalita et al. 2004; Acconcia and Kumar 2006). The biological effects are mediated by estrogens through their association with estrogen receptors (ERs) α and β. The ERα was first isolated by Jensen and Jacobson (1962). The discovery of ERβ by the Gustafsson's laboratory in 1996 led to better understanding of the pleiotropic effects of estrogens, even in tissue and cells lacking ERα (Acconcia and Kumar 2006). ERα and ERβ are encoded by various genes, which are located on different chromosomes. However, both have common features with other nuclear receptors. They are composed of five or six domains, which are labeled from A/B to F (Fig. 6.1). The A/B domains are situated on the amino terminus of the protein and contain the activation function 1 (AF-1) motif, which is responsible for activation of transcription independently from ligand binding. The C domain is involved in receptor dimerization and binding of ligand – receptor complex to specific DNA sequences. The D domain has also partially DNA-binding properties due to the nuclear localization signal (NLS). The carboxyl terminus of the protein attaches to the specific ligand. Some ERs contain an additional E domain, whose function is not well known (Simoncini et al. 2004).

Activation of these receptors involves nuclear translocation and binding to specific response elements, which control gene expression. This modulation of gene transcription by specific nuclear estrogen receptor (nER) was termed a "genomic" activity of estrogen in contrast to "non-genomic" effects of estrogen. The classic genomic estrogen mechanism requires binding of estrogens to the ER and transloca-

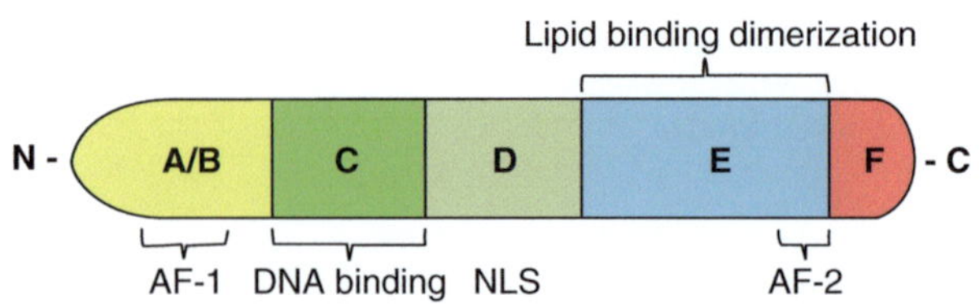

Fig. 6.1 Structure of the estrogen receptor; A/B domain with AF-1 motif (regulation function), C domain with DNA-binding domain, D domain with nuclear localization signal (*NLS*), E domain with ligand-binding domain and AF-2 motif, F domain (unknown function) (Świtalska and Strządała 2007)

tion to the nucleus. Subsequently, the dimerization of receptors is required which can result in the formation of homodimers or heterodimers (α-α and β-β or α-β). After dimerization, ER binds to specific DNA estrogen response elements (EREs) through its two-zinc finger motif (Felty and Roy 2005; Świtalska and Strządała 2007). EREs are located in various gene promoters. Estrogen binding can induce other modifications of the receptor conformation in the ligand-binding domain, such as association with coactivator protein. Many genes, which are regulated by estrogen, do not contain EREs, and the molecular mechanism through which estrogens regulate their transcription is not fully understood (Björnström and Sjoberg 2005). There is increasing evidence that estrogens may evoke their effects via various nonclassical signaling pathways and that ERs can regulate gene expression by several distinct mechanisms. Numerous investigations have shown that ERs might regulate transcription without binding directly to DNA. In such cases, the receptors are linked through protein-protein interactions to a transcription factor complex and, thus, contact the DNA. By means of this mechanism, ERs can regulate the expression of many estrogen-responsive genes that do not contain ERE. Many estrogen-regulated genes with the lack of EREs contain binding sites for an orphan nuclear hormone receptor SF-1 and SF-1 response element (SFRE), which determine ERα binding affinity (Ordonez-Moran and Munoz 2009; Levin and Hames 2016). Several effects induced by estrogens are too rapid to involve activation of RNA and protein synthesis. These observations have stimulated the search for alternative signaling mechanisms. Many investigations show that numerous crucial biological processes are not mediated by the classical estrogen hormone receptor action in the nucleus. Various terms have been used to differentiate these non-conventional receptor actions. In contrast to the classical genomic action, the non-genomic effects are characterized by a rapid onset of the action which ranges from few seconds to 1 min after exposure and indicates posttranslational modifications of signaling proteins. Numerous studies have shown that the majority of ERs are found in extranuclear locations including the plasma membrane. Through these, membrane estrogen receptors initiate fast signal activation, which affects various important parts of cellular biology (Simoncini et al. 2004; Contrò et al. 2015). This review summarizes the recent knowledge on the signaling pathways of ER, which are localized in the cell membrane and are activated by estrogen ligands.

6.2 Rapid, Non-transcriptional Regulatory Mechanisms

Until the beginning of the twentieth century, biological regulation of transcription was considered as the only action mechanism attributed to estrogen. Groundbreaking experiments conducted by Szego and Davis (1967) raised reasonable doubts about this prevailing view, since the authors observed a rapid interaction between hormones and their targets (Szego and Davis 1967).In addition, ample evidence was provided for the existence of non-genomic mechanisms, including the action upon cells which lack nuclei or contain tightly packed chromatin, actions in the presence of transcription inhibitors, or actions upon cells with mutated receptors, disabling the transcription (Falkenstein et al. 2000; Zielniok et al. 2014). The rapid effects of steroid hormones are manifold, ranging from activation of protein and tyrosine kinases and G-proteins and modulation of ion channels. There are few different ways how steroids may interact with the cell membrane. They may change membrane fluidity, without binding to any known protein or receptor. Another way is allosteric modulation of (1) receptors which are nonspecific for steroid hormones or (2) structural and enzymatic proteins present in the cell membrane (Lachowicz-Ochędalska 2005). Finally, membrane-bound ER may control intracellular signal transduction cascades. It is also possible that the different mechanism may interact.

6.2.1 Membrane Receptors Specific to Estrogens

There is more than sufficient evidence that ERα and ERß are both found in the cell membrane. These two membrane-associated receptors may occur either in the form of homo- (ERα-ERα or ERβ-ERβ) or hetero- (ERα-ERβ) dimers (Boonyaratanakornkit and Edwards 2007). According to current knowledge, the membrane estrogen receptor (mER) and the nuclear estrogen receptor (nER) do not differ because spectra, weight, electrophoretic mobility, and affinity to E2 are virtually identical in mER and nER (Soltysik and Czekaj 2013). Moreover, investigations with monoclonal antibodies indicate the similarity between the protein epitopes in these two receptor types (Pappas et al. 1995; Norfleet et al. 1999). However, several factors can affect the immune detection of mERs. Latest studies on the mERα identified several negative regulators of its expression such as serum starvation, lack of cell cycle synchronization within cell population, too high cell density, or too many cell passages (Watson et al. 2002). Furthermore, integrity of the cell membrane, flexibility of the epitopes, and interactions with other molecules should also be taken into consideration when developing immune-labeling protocols (Watson et al. 2002).

In addition to mERα and mERβ, few other proteins were classified as mER within the last two decades. Among them, the most investigated is the G-protein-coupled receptor 30 (GPER or Gpr30), a rhodopsin-like protein unrelated to steroid nuclear receptors. Recently, it has been reported that Gpr30 is not only involved in rapid regulatory mechanisms such as mobilization of calcium or kinase activation but also may play a significant role in transcriptional activation of genes such as c-fos (Prossnitz and Maggiolini 2009). Another receptor, functionally distinct form ERα and ERβ, was identified by Toran-Allerand et al. (2002) and was named ER-X. ER-X can be distinguished from previously described receptors inter alia by its molecular weight which is 63 kDa, while ERα, ERβ, and Gpr30 amount 67 kDa, 60 kDa, and 44 kDa, respectively (Almey et al. 2015). Moreover, it does not stimulate extracellular signal-regulated kinases (ERKs) in the presence of selective ER agonists and reacts more strongly to 17α than 17β estradiol (Soltysik and Czekaj 2013). In 2012, Kampa et al. discovered another receptor in T47D and MCF-7 cells which is associated with cell signaling, apoptosis, and transcriptional regulation (Kampa et al. 2012). This receptor was confusingly called ERx although there is no evidence that ER-X and Erx are related. However, both receptors are not yet examined thoroughly.

Although details of the association of ERs with the membrane remain unclear, two main interactions are believed to be crucial: the palmitoylation of these ERs and their interaction with membrane/cytoplasmic scaffolding proteins (Boonyaratanakornkit 2011). Palmitoylation is the process of posttranslational lipid modification which increases receptor hydrophobicity and, therefore, enables its anchoring in the membrane. The second interaction apparently occurs between receptors and certain proteins, especially caveolin – integral membrane proteins placed within the area of the plasma membrane invaginations called caveolae. Caveolae are specialized lipid-ordered domains which comprise numerous signaling molecules, especially kinases. Although most studies concern only mERα and mERβ, the association between caveolin-rich light membranes complexes and mERs was also emphasized for ER-X as a way of rapid interaction with the mitogen-activated protein kinase (MAPK) cascade and other signaling pathways (Toran-Allerand et al. 2002). The mode of mER action has not yet been fully understood. However, it is acknowledged that internalization and dimerization play a significant role in the regulation of this process. Since activation follows the dimerization of nERs, it is not surprising that a similar process occurs in their membrane-associated equivalents. Generally, mERs are assumed to be a multiprotein complex that requires the cooperation with many other molecules including insulin-like growth factor 1 receptor (IGFR), epidermal growth factor receptor (EGFR), and proteins involved in the MAPK pathway such as Ras protein and adaptor protein Shc (Soltysik and Czekaj 2013). The exact mechanism of mERα action was identified by Li et al. in human endothelial cells (Li et al. 2003). It has been found that after E2 stimulation, palmitoylated mER dimers activate endothelial nitric oxide synthase (eNOS) in cooperation with c-Src, PI-3 K, Akt kinase, and heat shock protein Hsp90, which results in additional production of nitric oxide (Li et al. 2003). Alternatively, E2 may induce the dissociation of G-proteins into two subunits: Gα and Gβ (Soltysik and Czekaj 2013). Noteworthy, the presented mechanism concerns only mERα. Thus, a complete understanding of the mechanisms of estrogen action requires more extensive research considering the specific nuclear and membrane estrogen receptors and the nonspecific interactions.

6.2.2 *Cell Membrane Ion Channels*

Estrogens can change membrane fluidity without direct binding to receptor proteins. The lipophilic hormone molecules can be incorporated into the cell membrane structures and may modify the function of membrane proteins and modulate ion channels. Estrogens can affect ion channels either directly or via some signaling pathways (Wróbel and Gregoraszczuk 2015). Potassium channels are the most investigated, followed by calcium, ligand-gated, sodium, and lastly chloride channels. Notably, the estrogen effects were not consistent with regard to most channel types. Thus, estrogens may elicit stimulatory effects, such as increasing channel activity or channel opening, but they may also provoke inhibitory effects such as decreasing, blocking, or eliminating the channel activity or opening probability. Steroids influence BK (big conductance Ca^{2+}- and voltage-activated K^{+}) and Ca-L

(L-type Ca^{2+}) channels. These diverse and conflicting effects do not speak in favor of a defined mechanism for rapid action (Lee-Ming and Pfaff 2016). For example, in vascular smooth muscle cells (VSMCs), E2 treatment inhibits the voltage-dependent L-type Ca^{2+} channels. E2 also controls K^+ efflux in VSMCs, by opening Ca^{2+}- and voltage-activated K^+ (BKCa) channels via cGMP-dependent phosphorylation. It is still debated whether the ER has any direct role in the control of membrane ion channels, but it is obvious that E2 can independently bind and activate Maxi-K^+ channels. These structures consist of a pore-forming α subunit and of a regulatory β subunit which increases the channel sensitivity to Ca^{2+}. E2 directly binds to the regulatory subunit, therefore activating the Maxi-K^+ channel (Simoncini and Genazzani 2003).

Some reports revealed that sex steroid hormones rapidly cause the inhibition of various cardiac ion currents. For example, they influence L-type Ca^{2+} currents, T-type Ca^{2+} currents, and slowly and rapidly activating delayed rectifier K^+ currents in isolated cardiomyocytes. In contrast to progesterone, 17β-estradiol exhibits such non-genomic action only at concentrations which are higher than the physiological serum concentrations (higher than 10 μM), suggesting only a minimal impact on the regulation of L-type Ca^{2+} and rectifier K^+ currents (Kurokawa and Furukawa 2013). Progesterone can decrease the membrane fluidity, aggregate membrane vesicles, and induce fusion of membrane vesicles. In contrast, testosterone and E2 at the same concentration had an insignificant effect on membrane fluidity, membrane aggregation, fusion, and leakage (Schmidt et al. 2000).

6.2.3 G-Protein-Coupled Receptors (GPCRs)

One of the best-known routes of non-genomic actions of sex steroids is the activation of receptors coupled to G-proteins. This transmembrane receptor is capable of binding to estrogen and its derivatives and thereby contributes to the control of both – physiological and pathological processes (Revankar et al. 2005). Filardo et al. (2000) showed that the fast response to E2 action results from the stimulation of extracellular kinase. This stimulation was shown to depend on the transmembrane orphan receptor GPR30 which was identified in the mid-1990s (O'Dowd et al. 1998), but not on ERα and ERβ. Based on gene's sequence homology and the mechanism of action, the GPR30 was classified as a member of the first superfamily of receptors, which contain seven transmembrane domains and are associated with G-proteins. The GPR30 is also abbreviated as GPER (Filardo et al. 2000). Through a rapid nonnuclear mechanism, this estrogen receptor is involved in the regulation of numerous physiological and pathological functions and often elicits biological effects that are opposite to the classical non-genomic action. GPER plays a major role in bone mineralization, energy control, and blood pressure regulation. GPER is also involved in immune responses and neuroprotection (Olde and Munoz 2009).

Modulation of mER linked to G-proteins may possibly explain a variety of cellular actions of sex steroids, such as regulation of the PLC/DAG/IP3 cascade, intracellular Ca^{2+} mobilization, activation of protein kinase C (PKC), and activation of the adenylate cyclase/protein kinase A pathway (Simoncini et al. 2003).

It was demonstrated that ERs can couple to isoenzyme β of phospholipase C (PLC) by interacting with G-protein in osteoblast cells. This leads to a rapid increase in intracellular Ca^{2+} concentration due to Ca^{2+} mobilization from the endoplasmic reticulum and to the formation of inositol 1,4,5-trisphosphate (IP3) and diacylglycerol (DAG) (Estrada et al. 2000; Simoncini and Genazzani 2003). The binding of the steroid hormone to its receptor results in dissociation of the G-protein into Gα and Gβγ subunits and activation of the kinase cascade (Soltysik and Czekaj 2013).

Up to now, the localization of GPER is not clearly defined. There are numerous studies indicating that this receptor is present in various cytoplasmic organelles. Whether GPER is located in the membrane, cytoplasm, or nucleus depends mainly on the type of cells (Cheng et al. 2014). The biological effects of GPER seem to depend on its level in the cell membrane (Mo et al. 2013).

6.2.4 Protein Kinase Signaling Pathways

The fast non-genomic effects of sex steroids are associated with the activation of signaling pathways in the cell which mediate the transmission of extracellular signals. These pathways involve MAPKs and tyrosine kinases (Simoncini and Genazzani 2003). The activation of the MAPK pathways by sex steroids has been characterized in several tissues. Three main cascades are ERK1/ERK2, the p38, and the stress-activated protein kinase (SAPK) or c-Jun NH2-terminal kinase (JNK) (Pearson et al. 2001).

Activation of MAPK may be controlled by the coordinated interaction with other signaling cascades. In breast cancer cells, E2-induced ERK activation is mediated by a heregulin (HRG)/human epidermal growth factor receptor-2/PKC-d/Ras pathway and leads to E2-dependent growth-promoting effects. Moreover, estrogen deprivation affects the breast cancer cells causing the development of hypersensitivity to the mitogenic effect of E2. There is some evidence that hypersensitivity to E2 depends on enhanced MAPK activity. These results might suggest a strong connection between the ER and the MAPK pathways (Yue et al. 2002).

The activation of mER induces the MAPK/ERK pathway. Considering recent reports, MAPK induction does not occur directly, but rather through the transactivation of EGFR. Numerous processes involved in proliferation are controlled by ERK. E2 promotes DNA synthesis in MCF-7 cells, visualized by the degree of radiolabeled thymidine incorporation. Application of MAPK inhibitor (PD98059) leads to 60–85% reduction of replication, indicating a significant contribution of the described pathways to the hormone-dependent progression of breast cancer (Soltysik and Czekaj 2013). On the other hand, E2-induced c-Src activation results in parallel activation of the MAP kinase (ERK1/ERK2) and Akt pathways, thereby resulting in eNOS phosphorylation and augmented NO release (Haynes et al. 2003). What is the significance of these pathways affected by sex steroids? The answer to this question seems to arise from the function of the tyrosine kinase SRC. Src, the gene encoding the tyrosine kinase protein SRC, is classified as a proto-oncogene; therefore, it takes part in the regulation of cell proliferation, adhesion, movement, and invasion, all of which are fundamental cellular processes. It is thus not surprising that the increase in its activity caused by mutations or by signals induced by a growth factor is observed in several types of cancer. Therefore, signaling via sex steroids, which activates the SRC/Ras/ERK pathway, would be engaged in stimulation of the cell proliferation (Migliaccio et al. 2007).

6.3 The Integration of Genomic and Non-genomic Estrogen Actions on the Phosphorylation of Transcription Factors

The classic action of estrogen depends on the nER. After combining with an estrogen, these receptors undergo phosphorylation followed by a regulation of the transcription of genes containing ERE in the promoter region (Davis et al. 2011). The nERs operate as ligand-activated transcription factors. However, as mentioned above, estrogen (E) may also act through unique mER (Vrtacnik et al. 2014).

Recent studies on estrogen mechanism of action provide evidence that non-genomic and genomic effects may be integrated into a unique mode of action (Pedram et al. 2014). An example for the convergence of estrogen genomic and non-genomic effects is the potential possibility to modulate gene expression. Lin et al. (2013) showed that isoform ER66 and ER46 displayed differential binding affinities with various agonists and antagonists and, in this way, regulate transcriptional activity. The study by Cato et al. (2002) revealed phosphorylation of ER46 by MAP kinase, thereby facilitating the binding of cofactors. On the other hand, Yamakawa and Arita (2004) showed that, in primary cultures, the inhibition of the MAPK activity is followed by suppression of E2-induced proliferation. E2-induced ERK activation upregulates transcription factor activator protein-1 (AP-1) which induces c-fos gene, a known coregulator for nuclear receptor action.

Ligands typical for nuclear receptors contribute to non-genomic signaling pathways, which often also elicit effects at the genomic level. Target genes of the pathway that involves mER are often activated independently of the nER. This non-genomic way of signaling is an additional mechanism by which ER may regulate transcription via phosphorylation of transcription factors (Pedram et al. 2014). It has been estimated that more than 500 kinases are encoded within the human genome. The diversity of signal transduction cascades to the nucleus unquestionably provides highly differentiated control processes of cellular functions. The signal initiated by the ERE contributes to cell proliferation and viability, affecting the functioning of both normal and neoplastic cells (Marino et al. 2006).

mER-mediated signals, such as cAMP-dependent protein kinase A (PKA) or MAPK, can target classical nERs and their coactivators. It can be said that mER, through kinase cascades, may affect and modulate nER action. Different kinases induced by mER phosphorylate several N-terminal residues of nER and thereby change receptor functions and reduce the transcriptional activity (Yamakawa and Arita 2004). Rapid phosphorylation of ERs causes their dimerization and nuclear translocation (Marino et al. 2006). For example, in breast cancer, quick E-mediated activation PI3K/Akt causes phosphorylation of Ser167 residue on ERα (Riggio et al. 2012), while phosphorylation induced by p38/MAPK or Thr311 blocks the ERα nuclear export and promotes the interaction of ERα with p160 steroid receptor coactivators (Lee and Bai 2002). Furthermore, phosphorylation of ERα (on Ser305) by TBK1 augments cyclin D1 transcription (Wei et al. 2014). Furthermore, cytoplasmic localization of ERα enhances the non-genomic effects, which has been proposed to contribute to tumorigenesis and anti-estrogen resistance of breast cancer cells (Vadlamudi et al. 2005; Gururaj et al. 2006).Constitutive activation of MAP kinase resulted in nuclear translocation of ERα. Lu et al. (2002) demonstrated membrane localization of transiently transfected ERα in human VSMC and translocation of ERα from the membrane to the nucleus in response to both estrogen-dependent and estrogen-independent stimulation. Authors showed that the pharmacological inhibition of MAP kinase activation blocked nuclear translocation of the ERα (Lu et al. 2002). Some observations suggest, however, that non-genomic signaling is necessary for classical nuclear signaling in vivo (Pedram et al. 2014). It was confirmed in vivo by a transgenic mouse model, in which the extranuclear signal was blocked. These mice showed the characteristics of the global ER knockout mouse with infertility, reduced mammary size, and other signs of hypogonadism (Hammes and Davis 2015). Currently, it is believed that these membrane and nuclear mechanisms are linear – starting with kinase signal and ending with changes in transcription. Rapid estrogen action can augment slower transcriptional processes mediated by nER. Moreover, induction of MAPK and PI3K pathway is necessary and sufficient not only for the transcription of non-nER-regulated genes but also for nER-regulated genes (Zhang and Trudeau 2006).

By the non-genomic action, the estrogens have an impact on proteins that keep nuclear receptors in the inactive state (heat shock protein) or on proteins which are responsible for the movement of classical receptors from the cytoplasm to the nucleus. In the absence of the estrogen ligand, the ERs are found in the nucleus within a heat shock protein complex that inhibits their action. The heat shock proteins displace upon binding to hormone, and this facilitates the binding of cofactors, which promote (coactivators) or inhibit (corepressors) the interaction of the receptors with their target genes (Echeverria and Picard 2010). It is very likely that estrogen, when reaching the cell, initiates a series of changes at the level of the cell membrane, which could be needed to enhance/enable the action of the newly synthesized proteins.

Further examples of multiple targets of ER action are cyclin D1 and vascular endothelial growth factor (VEGF) genes (Ordonez-Moran and Munoz 2009). Cyclin D1, a well-defined target for E2 in the mammary gland, is important in the cell division by controlling the progression of cells from the G1 to S phase of the cell cycle. It is mainly observed in cancer cells rather than in normal ones, and it is involved in the proliferation of tumor cells. The gene encoding cyclin D1 (CCND1) contains binding sites for numerous transcription factors, but no ERE-like sequences, suggesting that it may be controlled by anon-nER mechanism (Zhang and Trudeau 2006).

Finally, main signaling modes may cross each other and intermingle. ER-dependent enzymatic mechanisms and transcriptional control can either activate or inhibit these pathways. Certain transcription factors (e.g., AP-1, Elk, or cAMP response element-binding protein – CREB) are downstream targets of kinases and can be rapidly activated by phosphorylation. This activation transfers the non-genomic action to genomic effects (Watson et al. 2011). Therefore, separation into genomic and non-genomic responses/mechanisms becomes somewhat inaccurate. One of the most studied transcription factor, which is rapidly activated by E, is CREB. In many cell types (hippocampal or colonic carcinoma cells), CREB activates the induction of MAPK pathway, independently of the PKA pathway. Such activation of CREB induces expression of several genes, e.g., c-fos, and uncoupling protein-2 (Marino et al. 2006).

The impact of ER on gene transcription occurs in several separate ways, which are presented in the Fig. 6.2. The complexity of the mechanisms activated by ER suggests a greater influence of non-genomic modes and rapid signals on cellular molecular events. The integration of all these complex events is required for a complete cellular response.

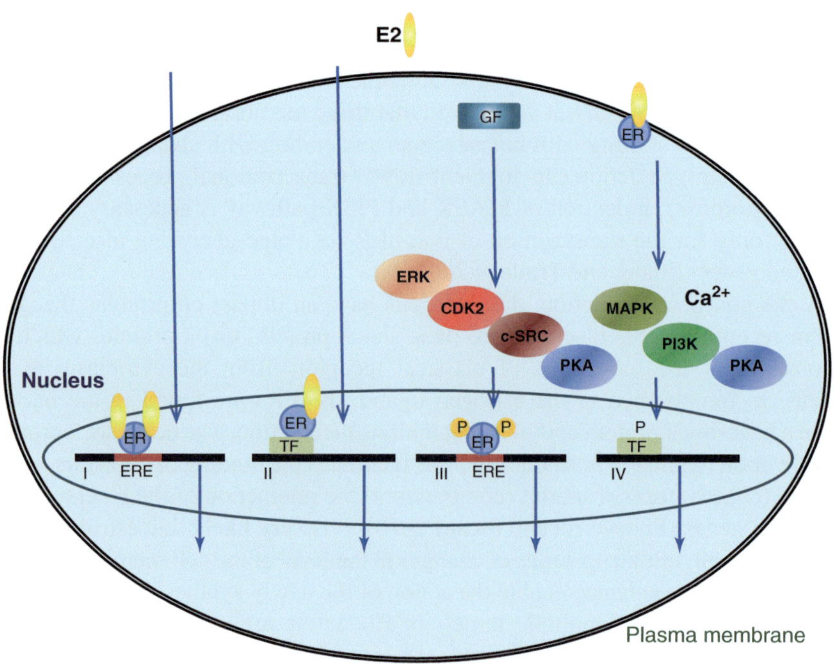

Fig. 6.2 Interactions of estrogen receptor with DNA. *I* classical mechanism, E2-ER complex binds directly to EREs in target gene promoters. *II* non-direct DNA-binding mechanism, ERE-independent genomic action and protein-protein interactions with other transcription factors as AP-1 or Sp-1. *III* ligand-independent genomic action, growth factors activate protein kinase cascades leading to phosphorylation (*P*) of ER at EREs. *IV* non-genomic mechanism, membrane-associated ERs mediate estrogen actions. *GF* growth factor, *TF* transcription factor, *P* phosphate, other abbreviations explained in the text

6.4 Summary

Apart from the influence on the formation of secondary sexual characteristics, estrogens have an impact on various cell processes. They regulate proliferation, growth, migration, and apoptosis of cells. Estrogen effects are elicited through genomic and non-genomic mechanisms, which are often studied separately. However, recent evidence suggests that the classical genomic and the non-genomic way of estrogen could be integrated by the rapid interaction of estrogen receptors with specific binding partners. The non-genomic effects of estrogens are complex and not yet fully elucidated. In order to fully understand estrogen's action, the effects of mER and the interaction between mER- and nER-dependent signaling need to be clarified.

Acknowledgments This work was supported by Statutory Funds of Wroclaw Medical University No.: ST.E130.17.013.

References

Acconcia F, Kumar R (2006) Signaling regulation of genomic and nongenomic functions of estrogen receptors. Cancer Lett 238:1–14

Almey A, Milner TA, Brake WG (2015) Estrogen receptors in the central nervous system and their implication for dopamine-dependent cognition in females. Horm Behav 74:125–138

Björnström L, Sjoberg M (2005) Mechanisms of estrogen receptor signaling: convergence of genomic and nongenomic actions on target genes. Mol Endocrinol 19(4):833–842

Boonyaratanakornkit V (2011) Scaffolding proteins mediating membrane-initiated extra-nuclear actions of estrogen receptor. Steroids 76:877–884

Boonyaratanakornkit V, Edwards DP (2007) Receptor mechanisms mediating non-genomic actions of sex steroids. Semin Reprod Med 25:139–153

Cato AC, Nestl A, Mink S (2002) Rapid actions of steroid receptors in cellular signaling pathways. Sci STKE 138:re9

Cheng SB, Dong J, Pang Y, La Rocca J, Hixon M, Thomas P, Filardo EJ (2014) Anatomical localization and redistribution of G protein -coupled estrogen receptor-1 during the estrus cycle in mouse kidney and specific bindings to estrogen but not aldosterone. Mol Cell Endocrinol 382:950–959

Contrò V, Basile JR, Proia P (2015) Sex steroid hormone receptors, their ligands, and nuclear and non-nuclear pathways. AIMS Mol Sci 2:294–310

Davis PJ, Lin HY, Mousa SA, Luidens MK, Hercbergs AA, Wehling M, Davis FB (2011) Overlapping nongenomic and genomic actions of thyroid hormone and steroids. Steroids 76(9):829–833

Echeverria PC, Picard D (2010) Molecular chaperones, essential partners of steroid hormone receptors for activity and mobility. BBA-Mol Cell Res 1803(6):641–649

Estrada M, Liberona JL, Miranda M, Jaimovich E (2000) Aldosterone- and testosterone –mediated intracellular calcium response in skeletal muscle cell cultures. Am J Physiol Endocrinol Metab 279:132–139

Falkenstein E, Tillmann HC, Christ M, Feuring MM, Wehling M (2000) Multiple actions of steroid hormones – a focus on rapid, nongenomic effects. Pharmacol Rev 52:513–555

Felty Q, Roy D (2005) Estrogen, mitochondria and growth of cancer and non-cancer cells. J Carcinog 4(1):1

Filardo EJ, Quinn JA, Bland KI, Frackelton AR Jr (2000) Estrogen-induces activation of Erk-1 and Erk-2 requires the G protein-coupled receptor homolog, GPR30, and occurs via transactivation of the epidermal growth factor receptor through release of HB-EGF. Mol Endocrinol 14:1649–1660

Gururaj AE, Rayala SK, Vadlamudi RK, Kumar R (2006) Novel mechanisms of resistance to endocrine therapy: genomic and nongenomic considerations. Clin Cancer Res 12(3):1001S–1007S

Hammes SR, Davis PJ (2015) Overlapping nongenomic and genomic actions of thyroid hormone and steroids. Best Pract Res Clin Endocrinol Metab 29(4):581–593

Haynes MP, Li L, Sinha D, Russell KS, Hisamoto K, Baron R, Collinge M, Sessa WC, Bender JR (2003) Src kinase mediates phosphatidylinositol 3-kinase/Akt-dependent rapid endothelial nitric-oxide synthase activation by estrogen. J Biol Chem 278(4):2118–2123

Jensen EV, Jacobson HI (1962) Basic guides to the mechanism of estrogen action. Recent Prog Horm Res 18(4):387

Kalita K, Lewandowski S, Skrzypczak M, Szymczak S, Tkaczyk M, Kaczmarek L (2004) In: Nowak JZ, Zawilska JB (eds) Receptory estrogenowe. Receptory i mechanizmy przekazywania sygnału. Wydawnictwo Naukowe PWN, Warszawa, pp 604–616

Kampa M, Notas G, Pelekanou V, Troullinaki M, Andrianaki M, Azariadis K, Kampouri E, Lavrentaki K, Castanas E (2012) Early membrane initiated transcriptional effects of estrogens in breast cancer cells: first pharmacological evidence for a novel membrane estrogen receptor element (ERx). Steroids 77:959–967

Kurokawa J, Furukawa T (2013) Non-genomic action of sex steroid hormones and cardiac repolarization. Biol Pharm Bull 36(1):8–12

Lachowicz-Ochędalska (2005) Membrane receptors for estradiol – new way of biological action. Endokrynol Pol 56(3):322–326

Lee H, Bai W (2002) Regulation of estrogen receptor nuclear export by ligand-induced and p38-mediated receptor phosphorylation. Mol Cell Biol 22(16):5835–5845

Lee-Ming K, Pfaff D (2016) Rapid estrogen actions on ion channels: a survey in search for mechanism. Steroids 111:46–53

Levin ER, Hammes SR (2016) Nuclear receptors outside the nucleus extranuclear signalling by steroid receptors. Nat Rev Mol Cell Biol 2016:783–797

Li L, Haynes MP, Bender JR (2003) Plasma membrane localization and function of the estrogen receptor alpha variant (ER46) in human endothelial cells. Proc Natl Acad Sci U S A 100:4807–4812

Lin AH, Li RW, Ho EY, Leung GP, Leung SW, Vanhoutte PM, Man RY (2013) Differential ligand binding affinities of human estrogen receptor-alpha isoforms. PLoS One 8(4):e63199

Lu Q, Ebling H, Mittler J, Baur WE, Karas RH (2002) MAP kinase mediates growth factor-induced nuclear translocation of estrogen receptor alpha. FEBS Lett 516(1–3):1–8

Marino M, Galluzzo P, Ascenzi P (2006) Estrogen signaling multiple pathways to impact gene transcription. Curr Genomics 7(8):497–508

Migliaccio A, Castoria G, Auricchio F (2007) Src-dependent signalling pathway regulation by sex-steroid hormones: therapeutic implications. Int J Biochem Cell Biol 39:1343–1348

Mo Z, Liu M, Yang F, Luo H, Li Z, Tu G, Yang G (2013) GPR30 as an initiator of tamoxifen resistance in hormone-dependent breast cancer. Breast Cancer Res 15(6):R114

Norfleet AM, Thomas ML, Gametchu B, Watson CS (1999) Estrogen receptor-alpha detected on the plasma membrane of aldehyde-fixed GH(3)/B6/F10 rat pituitary tumor cells by enzyme-linked immunocytochemistry. Endocrinology 140:3805–3814

O'Dowd BF, Nguyen T, Marchese A, Cheng R, Lynch KR, Heng HH, George SR (1998) Discovery of three novel G-protein-coupled receptor genes. Genomics 47(2):310–313

Olde B, Munoz A (2009) GPR30/GPER1: searching for a role in estrogen physiology. Trens Endocrinol Metab 20:409–416

Ordonez-Moran P, Munoz A (2009) Nuclear receptors. Genomic and non-genomic effects converge. Cell Cycle 8(11):1675–1680

Pappas TC, Gametchu B, Watson CS (1995) Membrane estrogen-receptors identified by multiple antibody labeling and impeded-ligand binding. FASEB J 9:404–410

Pearson G, Robinson F, Beers Gibson T, Xu BE, Karandikar M, Cobb MH (2001) Mitogen – activated protein (MAP) kinase pathways: regulation and physiological functions. Endocr Rev 22:153–183
Pedram A, Razandi M, Lewis M, Hammes S, Levin ER (2014) Membrane-localized estrogen receptor α is required for normal organ development and function. Dev Cell 29(4):482–490
Prossnitz ER, Maggiolini M (2009) Mechanisms of estrogen signaling and gene expression via GPR30. Mol Cell Endocrinol 308:32–38
Revankar CM, Cimino DF, Sklar LA, Arterburn JB, Prossnitz ER (2005) A transmembrane intracellular estrogen receptor mediates rapid cell signaling. Science 307:1625–1630
Riggio M, Polo ML, Blaustein M, Colman-Lerner A, Lüthy I, Lanari C, Novaro V (2012) PI3K/AKT pathway regulates phosphorylation of steroid receptors, hormone independence and tumor differentiation in breast cancer. Carcinogenesis 33(3):509–518
Schmidt BMW, Gerdes D, Feuring M, Falkenstein E, Christ M, Wehling M (2000) Rapid, nongenomic steroid actions: a new age? Front Neuroendocrinol 21:57–94
Simoncini T, Genazzani AR (2003) Non-genomic actions of sex steroid hormones. Eur J Endocrinol 148:281–292
Simoncini T, Mannella P, Fornari L, Caruso A, Varone G, Genazzani AR (2004) Genomic and nongenomic effects of estrogens on endothelial cells. Steroids 69:537–542
Soltysik K, Czekaj P (2013) Membrane estrogen receptors – is it an alternative way of estrogen action? J Physiol Pharmacol 64:129–142
Świtalska M, Strządała L (2007) Niegenomowe działanie estrogenów. Postępy Hig Med Dośw 61:541–547
Szego CM, Davis JS (1967) Adenosine 3′,5′-monophosphate in rat uterus – acute elevation by estrogen. Proc Natl Acad Sci U S A 58(4):1711–1718
Toran-Allerand CD, Guan XP, MacLusky NJ, Horvath TL, Diano S, Singh M, Connolly ES, Nethrapalli IS, Tinnikov AA (2002) ER-X: a novel, plasma membrane-associated, putative estrogen receptor that is regulated during development and after ischemic brain injury. J Neurosci 22:8391–8401
Vadlamudi RK, Manavathi B, Balasenthil S, Nair SS, Yang Z, Sahin AA, Kumar R (2005) Functional implications of altered subcellular localization of PELP1 in breast cancer cells. Cancer Res 65(17):7724–7732
Vrtačnik P, Ostanek B, Mencej-Bedrač S, Marc J (2014) The many faces of estrogen signaling. Biochem Med 24(3):329–342
Watson CS, Campbell CH, Gametchu B (2002) The dynamic and elusive membrane estrogen receptor-alpha. Steroids 67:429–437
Watson CS, Jeng YJ, Guptarak J (2011) Endocrine disruption via estrogen receptors that participate in nongenomic signaling pathways. J Steroid Biochem Mol Biol 127(1–2):44–50
Wei C, Cao Y, Yang X, Zheng Z, Guan K, Wang Q, Tai Y, Zhang Y, Ma S, Cao Y, GeX XC, Li J, Yan H, Ling Y, Song T, Zhu L, Zhang B, Xu Q, Hu C, Bian XW, He X, Zhong H (2014) Elevated expression of TANK-binding kinase 1 enhances tamoxifen resistance in breast cancer. Proc Natl Acad Sci U S A 111(5):E601–E610
Wróbel AM, Gregoraszczuk EŁ (2015) Action of methyl-, propyl- and butylparaben on GPR30 gene and protein expression, cAMP levels and activation of ERK1/2 and PI3K/Akt signaling pathways in MCF-7 breast cancer cells and MCF-10A non-transformed breast epithelial cells. Toxicol Lett 238(2):110–116
Yamakawa K, Arita J (2004) Cross-talk between the estrogen receptor-, protein kinase A-, and mitogen-activated protein kinase-mediated signaling pathways in the regulation of lactotroph proliferation in primary culture. J Steroid Biochem Mol Biol 88(2):123–130
Yue W, Wang JP, Conaway M, Masamura S, Li Y, Santen RJ (2002) Activation of the MAPK pathway enhances sensitivity of MCF-7 breast cancer cells to the mitogenic effect of estradiol. Endocrinology 143(9):3221–3229
Zhang DP, Trudeau VL (2006) Integration of membrane and nuclear estrogen receptor signaling. Comp Biochem Physiol A Mol Integr Physiol 144(3):306–315
Zielniok K, Gajewska M, Motyl T (2014) Molecular actions of 17 beta-estradiol and progesterone and their relationship with cellular signaling pathways. Postepy Hig Med Dosw 68:777–792

Chapter 7
How Imaging Membrane and Cell Processes Involved in Electropermeabilization Can Improve Its Development in Cell Biology and in Clinics

Laure Gibot, Muriel Golzio, and Marie-Pierre Rols

Abstract Cell membranes can be transiently permeabilized under the application of electric pulses. This process, called electropermeabilization or electroporation, allows hydrophilic molecules, such as anticancer drugs and DNA, to enter into cells and tissues. The method is nowadays used in clinics to treat cancers. Vaccination and gene therapy are other fields of application of DNA electrotransfer. A description of the mechanisms can be assayed by using different complementary systems with increasing complexities (models of membranes, cells cultivated in 2D and 3D culture named spheroids, and tissues in living mice) and different microscopy tools to visualize the processes from single molecules to entire animals. Single-cell imaging experiments revealed that the uptake of molecules (nucleic acids, antitumor drugs) takes place in well-defined membrane regions and depends on their chemical and physical properties (size, charge). If small molecules freely cross the electropermeabilized membrane and have a free access to the cytoplasm, larger molecules, such as plasmid DNA, face physical barriers (plasma membrane, cytoplasm crowding, nuclear envelope) which reduce transfection efficiency and engender a complex mechanism of transfer. Gene electrotransfer indeed involves different steps that include the initial interaction with the membrane, its crossing, transport within the cytoplasm, and finally gene expression. In vivo, additional very important effects of electric pulses are present such as blood flow modifications. The full knowledge on the way molecules are transported across the electropermeabilized membranes and within tissues is mandatory to improve the efficacy and the safety of the electropermeabilization process both in cell biology and in clinics.

L. Gibot • M. Golzio • M.-P. Rols (✉)
Institut de Pharmacologie et de Biologie Structurale, Université de Toulouse, CNRS, UPS, 205 Route de Narbonne, BP 64182, F-31077 Toulouse, France
e-mail: marie-pierre.rols@ipbs.fr

J. Kulbacka, S. Satkauskas (eds.), *Transport Across Natural and Modified Biological Membranes and its Implications in Physiology and Therapy*, Advances in Anatomy, Embryology and Cell Biology 227,
DOI 10.1007/978-3-319-56895-9_7

7.1 Introduction

Electropermeabilization, also called electroporation (EP), is based on the reversible permeabilization of cell membranes, thus enabling the delivery of non-permeant or poorly permeant molecules into the cells. The use of electric field pulses to deliver therapeutic molecules including drugs, proteins, and nucleic acids in a wide range of cells and tissues has been largely developed over the last decades (Andre and Mir 2010; Yarmush et al. 2014; Rosazza et al. 2016). This physical method is nowadays used in clinics to treat cancers, a process named electrochemotherapy (ECT). ECT combines the local application of well-defined electric field pulses following the local or systemic injection of antitumor drugs such as bleomycin and cisplatin (Mir et al. 1998; Escoffre and Rols 2012). Although ECT is used in routine clinical practice for treatment of subcutaneous tumors in more than 140 clinics and hospitals throughout Europe, mechanisms involved are still not fully understood. Reversible permeabilization of the cell membrane is the basic mechanism of the antitumor effectiveness. But the fact that ECT fairly preserves healthy tissues is not completely explained, as well as the effects of EP on blood circulation and vessels permeability.

Besides ECT, vaccination and oncology gene therapy are also major fields of application of DNA electrotransfer (Chiarella et al. 2010; Sersa et al. 2015). Translation of preclinical studies into clinical trials in human and veterinary oncology has started (Cemazar et al. 2010; Heller and Heller 2010). The first phase I dose escalation trial of electroporation of plasmid encoding for interleukin 12 has been carried out in patients with metastatic melanoma and has shown encouraging results (Daud et al. 2008). But the safe and efficient use of this physical method for clinical purposes requires the knowledge of the mechanisms underlying the electropermeabilization phenomena. Despite the fact that the pioneering work on plasmid DNA electrotransfer in cells was initiated 35 years ago (Neumann et al. 1982), many of the mechanisms underlying DNA electrotransfer remain to be elucidated (Teissie et al. 2005) as the way tissues respond to the electric pulses (Kamensek et al. 2016). Even if in vitro electrotransfer is efficient in almost all cell lines, in vivo gene delivery and expression in tumors are usually not (Rols et al. 1998). It is still mandatory, for increasing gene transfer and expression, to increase our knowledge of the different processes occurring both in vitro and in vivo.

This review focuses on (1) what it is (still not) known about the processes of transport of molecules across the membranes submitted to electric pulses and (2) how this knowledge helps to define electric parameters for improving the efficacy and safety of the electropermeabilization process both in cell biology and in clinics. It will describe key experiments obtained by using convenient imaging tools to directly visualize the different processes of transport and this on different experimental models with increasing complexities from simple model of membranes to tissues (Fig. 7.1).

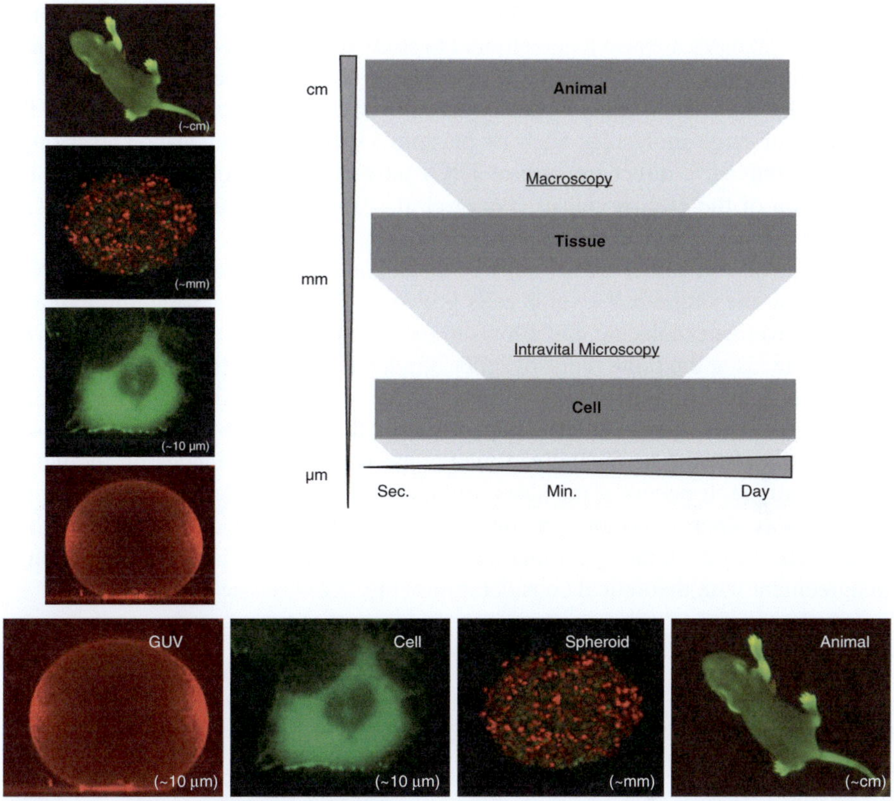

Fig. 7.1 Models of increasing complexities and imaging tools used to address electroporation processes

7.2 Imaging Membrane Processes

7.2.1 Membrane Electropermeabilization

Giant unilamellar vesicles (GUVs) are useful to study the effect of permeabilizing electric fields in simple membrane models. GUVs represent a convenient way to study membrane properties such as lipid bilayer composition and membrane tension. They offer the possibility to study and visualize membrane processes due to their cell-like size in absence of any constraint due to cytoskeleton. Experiments performed by phase contrast and fluorescence microscopies as well as by CARS spectroscopy showed a decrease in vesicle radius which is interpreted as being due to lipid loss during the permeabilization process (Mauroy et al. 2012). Three mechanisms responsible for lipid loss were directly observed, pore formation, vesicle formation, and tubule formation, which may be involved in molecule uptake (Portet et al. 2009). GUVs are also a good model to study the mechanisms of electrofusion, with a direct interest to their use as vehicles to deliver molecules (Mauroy et al. 2012). However, GUV cannot be used to study the whole process involved in the transport of large molecules such as plasmid DNA where nuclear expression is required. Indeed, a direct transfer of DNA into the GUVs took place during the application of the electric pulses (Portet et al. 2011), which is contradictory with what was observed in unique cell context.

The use of video microscopy allowed visualization of the permeabilization phenomenon at the single-cell level in cells grown on Petri dishes. Propidium iodide is a very convenient molecule that allows to monitor membrane electropermeabilization. Its size is at the same range of order of a large variety of therapeutic drugs. Its uptake into the cytoplasm is a fast process that is induced during electric pulse delivery and that can be detected seconds after the application of electric pulses. Exchange across the permeabilized membrane is not homogeneous on the whole cell membrane. It occurs at the sides of the cells facing the electrodes in an asymmetrical way where it is more pronounced at the anode-facing side of the cells than at the cathode, i.e., in the hyperpolarized area than in the depolarized area, which is in agreement with theoretical considerations (Fig. 7.2) (Teissie et al. 2005).

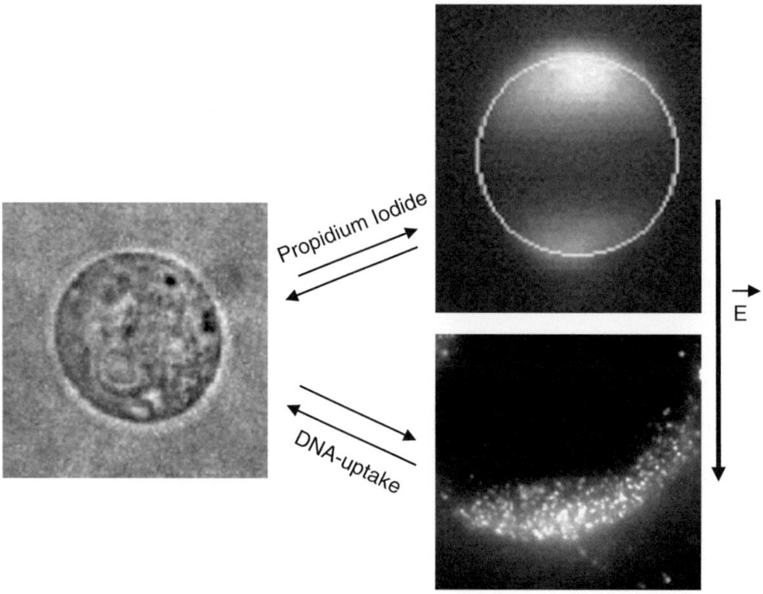

Fig. 7.2 Different steps involved in membrane electropermeabilization and gene electrotransfer as directly visualized under a microscope a few seconds following pulses delivery. Before pulse application, the membrane is used as a barrier that prevents the passage of small hydrophilic molecules such as propidium iodide and large and charged molecules such as plasmid DNA. Electric pulse application induces the permeabilization of the cell membrane facing the two electrodes and DNA interaction facing the cathode

Electropermeabilization can be described as a three-step process in respect with electric field: (1) before electropulsation, the plasma membrane acts as a physical barrier that prevents the free exchange of hydrophilic molecules between the cell cytoplasm and external medium; (2) during electropulsation, the transmembrane potential increases which induces the formation of transient permeable structures facing the electrodes and allows the exchange of molecules; and (3) after electropulsation, membrane resealing occurs within minutes.

A direct transfer into the cell cytoplasm of the negatively charged small molecules such as siRNA is observed on the side facing the cathode. When added after electropulsation, siRNA do not penetrate the cells. Therefore, electric field acts on both the permeabilization of the membrane and on the electrophoretic drag of the charged molecules from the bulk into the cytoplasm. The mechanism involved is clearly specific for the physicochemical properties of the electrotransferred molecule (Paganin-Gioanni et al. 2011; Golzio and Teissie 2014).

Progress in the knowledge of the involved mechanisms at the molecular level is still a biophysical challenge. Once again, fluorescence microscopy helped to go deeper into the elucidation of the mechanisms. The electric pulses induced the formation of long-lived permeant structures and resulted in a rapid phospholipid flip-flop within less than 1 s and were exclusively restricted to the regions of the permeabilized membrane. These results could support the existence of direct interactions between the movement of membrane zwitterionic phospholipids and the electric field (Escoffre et al. 2014). In addition, experiments on lateral mobility of proteins showed that electropermeabilization affects the lateral mobility of membrane protein, a result that suggests that 10–20% of the membrane surface is occupied by defects or pores and that these structures propagate rapidly over the cell surface (Escoffre et al. 2014).

It is also possible to take advantage of atomic force microscopy to directly visualize the consequences of electropermeabilization (without using any fluorescent molecule) and to locally measure the membrane elasticity. The transient rippling of membrane surface has been visualized which was associated to a decrease in membrane elasticity. These results obtained both on fixed and living cells give evidence of an inner effect affecting the entire cell surface which may be related to cytoskeleton destabilization and not only the area where transfer of molecules takes place (Chopinet et al. 2013).

7.2.2 *Electrotransfer of DNA Molecules*

Single-cell microscopy and fluorescent plasmids can be used to monitor the different steps of gene electrotransfer (GET) (Golzio et al. 2002). As in the case of siRNA, DNA molecules, which are negatively charged, migrate electrophoretically. Under electric fields that are too small to permeabilize the membrane, the DNA flows around the cell. Beyond a critical field value, above which cell permeabilization occurs, the DNA interacts with the electropermeabilized plasma membrane. This interaction only occurs at the pole of the cell opposite the cathode, and this demonstrates the importance of electrophoretic forces in the initial phase of the DNA/membrane interaction. DNA/membrane interaction is correlated to the formation of "microdomains" whose dimensions lie between 0.1 and 0.5 μm (Fig. 7.2). DNA electrotransfer can be described as a multistep process: the negatively charged DNA migrates electrophoretically toward the plasma membrane on the cathode side where it accumulates. This interaction, which is observed for several minutes, lasts much longer than the duration of the electric field pulses. Translocation of the plasmid from the plasma membrane to the cytoplasm and its subsequent passage toward the nuclear envelope take place with a kinetics ranging from minutes to hours. When plasmids have reached the nuclei, gene expression can take place, and this can be detected up to several days in the case of dividing cells or weeks in some tissues such as muscles.

The dynamic of this process has been poorly understood because direct observations have been limited to time scales that exceed several seconds. The use of a camera with a temporal resolution of 2 ms allowed the visualization of the DNA/membrane interaction process during pulse application. DNA molecules interact with the membrane during the application of the pulse. At the beginning of the pulse application, plasmid complexes or aggregates appear at specific sites on the cell membrane. The formation of plasmid complexes at fixed sites suggests that membrane domains may be responsible for DNA uptake and their lack of mobility could be due to their interaction with the actin cytoskeleton. FRAP measurements show that the positions of these sites are remarkably immobile during the application of further pluses. A theoretical model is proposed to explain the appearance of distinct interaction sites, the quantitative increase in DNA, and also their immobility leading to a tentative explanation for the success of electromediated gene delivery (Escoffre et al. 2011).

DNA/membrane interaction and gene expression depend on electric pulse polarity, repetition frequency, and duration. Both are affected by reversing the polarity and by increasing the repetition frequency or the duration of pulses. The results revealed the existence of two classes of DNA/membrane interaction: (1) a metastable DNA/membrane complex from which DNA can leave and return to external medium and (2) a stable DNA/membrane complex, where DNA cannot be removed, even by applying electric pulses of reversed polarity. Only DNA belonging to the second class leads to effective gene expression. The lifetime of DNA/membrane complex formation is in the order of 1 s and has to be taken into account to improve protocols of electromediated gene delivery (Faurie et al. 2010).

Even if the first stage of gene electrotransfection, i.e., migration of plasmid DNA toward the electropermeabilised plasma membrane and its interaction with it, becomes understood, it is not totally possible today to give guidelines to improve gene electrotransfer. Successful expression of the plasmid depends on its subsequent migration into the cell. Therefore, the intracellular diffusional properties of plasmid DNA, as well as its metabolic instability and nuclear translocation, represent other cell-limiting factors that must be taken into account. In the conditions induced during electropermeabilization, the time a plasmid DNA takes to reach the nuclei is significantly longer than the time needed for a small molecule (Rosazza et al. 2011). Single-particle tracking experiments of individual DNA aggregates in living cells allowed further elucidation of the mechanism of DNA transfer. Active transport was visualized over long distances and has been shown to be related to the cellular microtubule network (Rosazza et al. 2013). In addition, the use of inhibitors of endocytosis and endosomal markers showed that, during active transport, DNA is routed through endosomal compartments.

Clear limits of efficient gene expression using electric pulses are therefore due to the passage of DNA molecules through the plasma membrane and to the cytoplasmic crowding and transfer through the nuclear envelope. A key challenge for electromediated gene therapy is to pinpoint the rate-limiting steps in this complex process and to find strategies to overcome these obstacles. One of the possible strategies to enhance DNA uptake into cells and to induce effects that primarily affect intracellular structures and functions is to use short (10–300 ns) but high pulse (up to 300 kV/cm) (Beebe et al. 2003). The idea, to improve transfection success, is thus to perform classical membrane permeabilization allowing plasmid DNA electrotransfer to the cell cytoplasm and, thereafter, when DNA has reached the nuclear envelope, to specifically permeabilize the nucleus using these short strong nanopulses. Thus, when used in conjunction with classical electropermeabilization, nanopulses gave hope to increase gene expression. Another idea is to combine electric pulses and ultrasound assisted with gas microbubbles, known as sonoporation. Cells that received electrosonoporation demonstrated a fourfold increase in transfection level and a sixfold increase in transfection efficiency compared with cells that have undergone electroporation alone (Escoffre et al. 2010). Although electroporation induced the formation of DNA aggregates into the cell membrane, sonoporation induced its direct propulsion into the cytoplasm. Sonoporation can therefore improve the transfer of electro-induced DNA aggregates by allowing its free and rapid entrance into the cells. These results demonstrated that in vitro gene transfer by electrosonoporation could provide a new potent method for gene transfer.

7.3 Imaging Cells and Tissue Processes

7.3.1 3D Cell Culture Models

Historically the culture of mammalian cells for laboratory uses has been performed on Petri dishes. However, it is nowadays clear that 2D studies do not translate well to the 3D microenvironment. Over the last several decades, 2D and 3D tissue engineering approaches have been developed to better mimic the complex architecture and properties of in vivo tissue. Literally, a new dimension to research has been achieved by the advent of three-dimensional cell culture techniques to bridge the gap between the "absolute in vitro" and "true in vivo" (Ravi et al. 2016).

Therefore, in the last past few years, in vitro tissue models, namely, multicellular tumor spheroids and cell sheets, have been used for the understanding of the electrotransfer processes in tumors and skin.

In order to assess the effects of extracellular matrix (ECM) composition and organization as well as intercellular junctions in tissue response to electric pulses, 3D human dermal tissue was reconstructed in vitro by a tissue engineering approach named self-assembly (Madi et al. 2015). This human cell model presented multiple layers of primary human dermal fibroblasts embedded in a native, collagen-rich

ECM. Cells of the reconstructed cutaneous tissue were efficiently electropermeabilized by applying millisecond electric pulses, without affecting their viability. A reporter gene was successfully electrotransferred and gene expression was detected for up to 48 h. Interestingly, the transfected cells were solely located on the upper surface of the tissue, where they were in close contact with plasmid DNA solution. Furthermore, results provided evidence that electrotransfection success depends on plasmid mobility within tissue rich in collagens, but not on cell proliferation status. In addition to proposing a reliable alternative to animal experiments, tissue engineering produces valid biological tool for the in vitro study of gene electrotransfer mechanisms in human tissue. A better comprehension of gene electrotransfer in such a model tissue would help to improve electrogene therapy approaches such as the systemic delivery of therapeutic proteins and DNA vaccination.

In order to in vitro mimic tumors, multicellular spheroids have been developed (Sutherland 1988). Upon growth, spheroids display a gradient of nutrients, metabolites, and proliferating cells. These proliferating cells are located in the outer cell layers, and the quiescent cells are located more centrally. This cell heterogeneity is similar to that found in avascular micro-regions of tumors. Confocal microscopy permits to visualize the repartition of permeabilized cells in spheroids submitted to electric pulses and to unravel gene transfer mechanisms. Experimental results revealed that cells were efficiently permeabilized, whatever their localization in the spheroid, even those in the core. Electrotransfer of bleomycin and cisplatin confirmed the relevance of the model in the case of electrochemotherapy, and doxorubicin showed its potential to screen new antitumor drug candidates for ECT. The combination of antitumor drugs and electric pulses indeed led to changes in spheroid macroscopic morphology and cell cohesion, to tumor spheroid growth arrest, and finally to its complete dislocation, mimicking previously observed in vivo situations (Gibot et al. 2013). In addition, comparison of transfection efficiency between cells in suspension and cells in spheroid allowed highlighting fundamental differences with 2D cell cultures. Using this 3D spheroid cell culture model also allows to study the effect of calcium electroporation and electrochemotherapy using bleomycin on human cancer cell lines and on primary normal human dermal fibroblasts. The results showed a clear reduction in spheroid size in spheroids after treatment with, respectively, calcium electroporation or electrochemotherapy using bleomycin. Strikingly, the size of normal fibroblast spheroids was affected neither after calcium electroporation nor after electrochemotherapy indicating that calcium electroporation, like electrochemotherapy, will have limited adverse effects on the surrounding normal tissue when treating with calcium electroporation (Frandsen et al. 2015).

The spheroid model therefore allows to study and optimize electromediated drug delivery protocols (Gibot and Rols 2013). Small molecules can be efficiently transferred into cells, including the ones present inside the spheroids, but gene expression is limited to the external layers of cells. Taken together, these results are in agreement with the ones obtained by the group of R. Heller in the USA (Marrero and Heller 2012) and indicate that the spheroid model is more relevant to an in vivo situation than cells cultured as monolayers.

7.3.2 *Tissue Responses*

Besides permeabilization of membranes, another important effect of electroporation that occurs in tissues (and which cannot be observed in vitro) is the blood flow-modifying effect (i.e., vascular lock) (Gehl et al. 2002). Electric pulses indeed induce a blood volume reduction in tumors (Sersa et al. 1999) and a short-term reduction of perfusion in muscles (Gehl et al. 2002). It has been observed that EP induces blood-modifying effects in tumors and normal tissues, but when combined with chemotherapeutics, it also results in disruption of tumor vasculature, without affecting adjacent normal vasculature. This dual effect of electrochemotherapy has clinical importance. On the one hand, due to the vascular disrupting action of ECT, it has proven to be successful in the treatment of bleeding tumors, and on the other hand, it can be safely applied to tumors that lie near large normal blood vessels, as was recently demonstrated in a clinical study on liver metastases of colorectal cancer. Moreover, even when used close to the heart, electroporation proved to be a safe method.

Direct observation with intravital fluorescence digitized microscopy imaging allowed to directly visualize the effects of electric pulses on the subcutaneous blood vessel dynamics and molecule electrotransfer (Bellard et al. 2012). These features were measured in mice via a dorsal skinfold window chamber, using fluorescently labeled dextrans of different sizes. Application of electric pulses on the skin in vivo resulted in a rapid increase in vascular permeability that gradually recovered to basal levels at different times posttreatment, depending on dextran size. Simultaneously, the immediate constriction of the blood vessels occurred which was more pronounced for arterioles compared to venules. This vasoconstriction of arterioles results in a transient "vascular lock." The increased permeability of small vessel walls whatever the dextran size associated with delayed perfusion can explain the improved delivery of the intravenous injected molecules (i.e., drugs, gene delivery) into the tissues induced by electropermeabilization in vivo. The study showed that plasmid DNA is sensitive to vascular lock. Namely, constriction of vasculature delays movement of large-sized molecules through the vasculature wall. Therefore, caution is needed when DNA is administered intravenously.

By using noninvasive bioluminescence technology, it has been recently possible to further explore the phenomena associated with GET to tumors by a real-time monitoring of the transfection efficiency as well as cell death following the treatment. Results showed that the GET of a reporter gene can lead to nonspecific antitumor effectiveness and even complete regression of tumors. Additionally, using the intratumoral GET of a luciferase-encoding plasmid, the source of the expression was localized mainly in the peritumoral and not in the tumoral region. These data provide new insights into some of the phenomena associated with GET into tumors, which should be taken into account when designing improved and more effective cancer gene therapy, in order to accelerate the transfer of the technology into clinical trials (Kamensek et al. 2016).

7.4 Summary and Future Direction

Clinical delivery of cytotoxic drugs is nowadays successfully used to treat tumors in patients. GET was originally used as a laboratory tool to deliver DNA to bacterial and mammalian cells in culture. Pulse generators and electrode developments allowed the method to be successfully used for in vivo uses. Several clinical trials related to delivery of plasmid DNA are promising in both cancer therapeutic and infectious disease vaccine applications. One huge benefit of the electropermeabilization technique besides its safety, efficiency, and low cost is its malleability. Electric pulse protocols can be fine-tuned to control the location, levels, and duration of subsequent transgene expression (Heller and Heller 2015).

Efficient GET is dependent on several cell- and tissue-related factors including extracellular matrix density and composition, plasmid DNA uptake, and nucleocytoplasmic transport. Different barriers are encountered by plasmid DNA from the extracellular environment toward the interior of the cell, and different strategies must be developed to overcome these biological barriers. A better understanding of the cellular and molecular bases of the physical gene transfer process may provide strategies to overcome those obstacles that highly limit the efficiency and use of gene delivery methods (Escoffre et al. 2010). Therefore, it is necessary to develop and use different models, from simple lipid vesicles to tumor multicellular tumor spheroids closer to the in vivo situation, for the understanding of the membrane permeabilization and DNA electrotransfer process in tissues. Each of these models has advantage and limits. When combined they can help to study the complete process (Table 7.1).

We believe that a full comprehension of the process involved in electropermeabilization and gene electrotransfer would further improve therapeutic approaches.

Table 7.1 Models to address electropermeabilization and gene delivery processes

Model	Membrane permeabilization	DNA electrotransfer
GUV	Visualization of membrane effects (deformation, lipid loss)	Failed to mimic DNA/membrane interaction (DNA is directly transferred inside the vesicle)
2D cell culture	Consequences of membrane permeabilization on membrane organization (lateral and transverse mobility of lipids and proteins)	Visualization of DNA/membrane complex formation and DNA traffic into the cells
3D cell culture	Molecules transfer that fairly mimic in vivo complex situations (contacts between cells, junctions, extracellular matrix)	Allow to study DNA delivery in 3D and mimic in vivo situation (decrease in gene expression from the periphery to the core)
Small animal (window chamber)	Visualization of blood flow modifications (vascular lock)	Allow to address in vivo effects (DNA sensitivity to vascular lock and constriction of vasculature delaying movement of large-sized molecules through the vasculature wall)

References

Andre FM, Mir LM (2010) Nucleic acids electrotransfer in vivo: mechanisms and practical aspects. Curr Gene Ther 10(4):267–280

Beebe SJ, White J et al (2003) Diverse effects of nanosecond pulsed electric fields on cells and tissues. DNA Cell Biol 22(12):785–796

Bellard E, Markelc B et al (2012) Intravital microscopy at the single vessel level brings new insights of vascular modification mechanisms induced by electropermeabilization. J Control Release 163(3):396–403

Cemazar M, Jarm T et al (2010) Cancer electrogene therapy with interleukin-12. Curr Gene Ther 10(4):300–311

Chiarella P, Fazio VM et al (2010) Application of electroporation in DNA vaccination protocols. Curr Gene Ther 10(4):281–286

Chopinet L, Roduit C et al (2013) Destabilization induced by electropermeabilization analyzed by atomic force microscopy. Biochim Biophys Acta 1828(9):2223–2229

Daud AI, DeConti RC et al (2008) Phase I trial of interleukin-12 plasmid electroporation in patients with metastatic melanoma. J Clin Oncol 26(36):5896–5903

Escoffre JM, Rols MP (2012) Electrochemotherapy: progress and prospects. Curr Pharm Des 18:3406–3415

Escoffre JM, Teissie J et al (2010a) Gene transfer: how can the biological barriers be overcome? J Membr Biol 236(1):61–74

Escoffre JM, Kaddur K et al (2010b) In vitro gene transfer by electrosonoporation. Ultrasound Med Biol 36(10):1746–1755

Escoffre JM, Portet T et al (2011) Electromediated formation of DNA complexes with cell membranes and its consequences for gene delivery. Biochim Biophys Acta 1808(6):1538–1543

Escoffre JM, Bellard E et al (2014a) Membrane disorder and phospholipid scrambling in electropermeabilized and viable cells. Biochim Biophys Acta 1838(7):1701–1709

Escoffre JM, Hubert M et al (2014b) Evidence for electro-induced membrane defects assessed by lateral mobility measurement of a GPi anchored protein. Eur Biophys J 43:277–286

Faurie C, Rebersek M et al (2010) Electro-mediated gene transfer and expression are controlled by the life-time of DNA/membrane complex formation. J Gene Med 12(1):117–125

Frandsen SK, Gibot L et al (2015) Calcium electroporation: evidence for differential effects in normal and malignant cell lines, evaluated in a 3D spheroid model. PLoS One 10(12):e0144028

Gehl J, Skovsgaard T et al (2002) Vascular reactions to in vivo electroporation: characterization and consequences for drug and gene delivery. Biochim Biophys Acta 1569(1–3):51–58

Gibot L, Rols MP (2013) Progress and prospects: the use of 3D spheroid model as a relevant way to study and optimize DNA electrotransfer. Curr Gene Ther 13(3):175–181

Gibot L, Wasungu L et al (2013) Antitumor drug delivery in multicellular spheroids by electropermeabilization. J Control Release 167(2):138–147

Golzio M, Teissie J (2014) siRNA delivery via electropulsation: a review of the basic processes. Methods Mol Biol 1121:81–98

Golzio M, Teissie J et al (2002) Direct visualization at the single-cell level of electrically mediated gene delivery. Proc Natl Acad Sci U S A 99(3):1292–1297

Heller LC, Heller R (2010) Electroporation gene therapy preclinical and clinical trials for melanoma. Curr Gene Ther 10(4):312–317

Heller R, Heller LC (2015) Gene electrotransfer clinical trials. Adv Genet 89:235–262

Kamensek U, Rols MP et al (2016) Visualization of nonspecific antitumor effectiveness and vascular effects of gene electro-transfer to tumors. Curr Gene Ther 16(2):90–97

Madi M, Rols MP et al (2015) Efficient in vitro electropermeabilization of reconstructed human dermal tissue. J Membr Biol 248:903–908

Marrero B, Heller R (2012) The use of an in vitro 3D melanoma model to predict in vivo plasmid transfection using electroporation. Biomaterials 33(10):3036–3046

Mauroy C, Castagnos P et al (2012a) Interaction between GUVs and catanionic nanocontainers: new insight into spontaneous membrane fusion. Chem Commun (Camb) 48(53):6648–6650
Mauroy C, Portet T et al (2012b) Giant lipid vesicles under electric field pulses assessed by non invasive imaging. Bioelectrochemistry 87:253–259
Mir LM, Glass LF et al (1998) Effective treatment of cutaneous and subcutaneous malignant tumours by electrochemotherapy. Br J Cancer 77(12):2336–2342
Neumann E, Schaefer-Ridder M et al (1982) Gene transfer into mouse lyoma cells by electroporation in high electric fields. EMBO J 1(7):841–845
Paganin-Gioanni A, Bellard E et al (2011) Direct visualization at the single-cell level of siRNA electrotransfer into cancer cells. Proc Natl Acad Sci U S A 108(26):10443–10447
Portet T, Camps i Febrer F et al (2009) Visualization of membrane loss during the shrinkage of giant vesicles under electropulsation. Biophys J 96(10):4109–4121
Portet T, Favard C et al (2011) Insights into the mechanisms of electromediated gene delivery and application to the loading of giant vesicles with negatively charged macromolecules. Soft Matter 7(8):3872–3881
Ravi M, Ramesh A et al (2016) Contributions of 3D cell cultures for cancer research. J Cell Physiol 232(10): 2679–2697
Rols MP, Delteil C et al (1998) In vivo electrically mediated protein and gene transfer in murine melanoma. Nat Biotechnol 16(2):168–171
Rosazza C, Escoffre JM et al (2011) The actin cytoskeleton has an active role in the electrotransfer of plasmid DNA in mammalian cells. Mol Ther 19(5):913–921
Rosazza C, Buntz A et al (2013) Intracellular tracking of single plasmid DNA-particles after delivery by electroporation. Mol Ther 21:2217–2226
Rosazza C, Meglic SH et al (2016) Gene electrotransfer: a mechanistic perspective. Curr Gene Ther 16(2):98–129
Sersa G, Cemazar M et al (1999) Tumor blood flow modifying effect of electrochemotherapy with bleomycin. Anticancer Res 19(5B):4017–4022
Sersa G, Teissie J et al (2015) Electrochemotherapy of tumors as in situ vaccination boosted by immunogene electrotransfer. Cancer Immunol Immun 64:1315–1327
Sutherland RM (1988) Cell and environment interactions in tumor microregions: the multicell spheroid model. Science 240(4849):177–184
Teissie J, Golzio M et al (2005) Mechanisms of cell membrane electropermeabilization: a minireview of our present (lack of ?) knowledge. Biochim Biophys Acta 1724(3):270–280
Yarmush ML, Golberg A et al (2014) Electroporation-based technologies for medicine: principles, applications, and challenges. Annu Rev Biomed Eng 16:295–320

Chapter 8
Analysis, Recognition, and Classification of Biological Membrane Images

Marek Kulbacki, Jakub Segen, and Artur Bak

Abstract Biological membrane images contain a variety of objects and patterns, which convey information about the underlying biological structures and mechanisms. The field of image analysis includes methods of computation which convert features and objects identified in images into quantitative information about biological structures represented in these images. Microscopy images are complex, noisy, and full of artifacts and consequently require multiple image processing steps for the extraction of meaningful quantitative information. This review is focused on methods of analysis of images of cells and biological membranes such as detection, segmentation, classification and machine learning, registration, tracking, and visualization. These methods could make possible, for example, to automatically identify defects in the cell membrane which affect physiological processes. Detailed analysis of membrane images could facilitate understanding of the underlying physiological structures or help in the interpretation of biological experiments.

8.1 Introduction

Most of our knowledge about microscopic structures including cells and membranes is conveyed via images. Extensive amounts of cell image data can now be automatically acquired and collected from a variety of microscopy techniques and multiple spectral channels. Tools and techniques that can help organizing and making sense out of such vast image collections are provided mainly by the fields of image analysis and processing, computer vision, and machine learning. These tools allow the construction of automated or semiautomated systems for

M. Kulbacki (✉) • J. Segen • A. Bak
Polish-Japanese Academy of Information Technology, Warsaw, Poland
e-mail: mk@pja.edu.pl; js@pja.edu.pl; abak@pja.edu.pl

J. Kulbacka, S. Satkauskas (eds.), *Transport Across Natural and Modified Biological Membranes and its Implications in Physiology and Therapy*,
Advances in Anatomy, Embryology and Cell Biology 227,
DOI 10.1007/978-3-319-56895-9_8

comparing, categorizing, classifying, and tracking images of cells, nuclei, or biological membranes. They can help identifying the phases of cell cycle or distinguishing damaged from healthy cells.

One of the problems in cell image analysis is the inhomogeneity of cell shapes and sizes. The main reason is lack of reference and additional variety caused by proliferation, cycle phases, and pathology. Biological membranes as natural and visible boundaries of cell structures are good candidates for research on cell state, condition, shape, separation, and many other qualitative and quantitative studies.

Along with reading this chapter, it may be helpful to review the glossary (Roeder et al. 2012) of terms used in the computational analysis of images, written from a biological perspective. It is a compact survey of image analysis techniques including image formats and conversions, common image operations and preprocessing techniques, and goal-oriented methods such as segmentation or registration.

Section 8.2.1 presents basic image processing operations. It is followed by detection methods with examples in Sect. 8.2.2, segmentation methods in Sect. 8.2.3, classification methods in Sect. 8.2.4, registration methods in Sect. 8.2.5, and tracking methods in Sect. 8.2.6. Visualization methods are discussed in Sect. 8.2.7. The last Sect. 8.2.8 describes open-source software for biological image analysis.

8.2 Cell Membrane Analysis

8.2.1 Basic Image Processing Operations

The basic image processing operations are functions on images that are commonly used as components of more complex methods and imaging techniques oriented toward specific applications. These operations include basic array operators, image conversion, measurement, comparison, filtering, enhancement, transformations, and statistics. They are presented in a systematic way in image processing textbooks (Gonzalez and Woods 2017; Pratt 2007; Russ and Neal 2016) and with examples in specialized medical imaging literature (Glasbey and Horgan 1995; Haidekker 2011; Jan 2006; Sonka and Fitzpatrick 2009).

The largest group of basic operations consists of functions that apply the same operator or process to every image pixel or pixel's neighborhood (with adjustments near image borders). The single pixel operations include arithmetic, logical, and comparison operators applied to pixels of a single image such as multiplying each pixel by a constant or passing it to a gamma function and multiple image operators such as pixel-by-pixel addition of two images or average of a group of images. For example, a frequently used image thresholding operation applied to a gray-scale image compares each pixel to a constant value returning 0 or 1, and the result is a binary image.

The common operations on pixel neighborhoods are convolution, correlation, and specific linear or nonlinear filters, such as: moving average (neighborhood mean); neighborhood median, maximum, or minimum; Gaussian; Laplacian; Laplacian of Gaussian (LoG); difference of Gaussian (DoG); or zero-crossing operator.

Convolution is used as a general linear filter, specified by its kernel or mask. Correlation can be used to identify image locations where a pixel's *NxN* neighborhood is similar to a given *NxN* image.

A separate group of neighborhood operations are the functions of mathematical morphology. They include erosion, dilation, opening, and closing. Erosion may be thought of as removing outer layers of binary regions, which results in thinning of lines and shrinking of blobs. Dilation adds outer layers to binary regions, resulting in thickening of lines and expanding of regions. Opening operation consists of applying erosion followed by dilation, which results in removing small blobs ("salt noise"), smoothing region edges, and rounding outer corners. Closing operation applies dilation after erosion, which results in filling small gaps and holes.

Examples of image transforms are discrete Fourier transform (DFT), discrete cosine transform, Walsh and Hadamard transform, and wavelet transform. Common uses of the DFT are speeding up convolution and correlation operations, based on the convolution theorem and designing linear filters in frequency domain. In most applications, a fast implementation of the DFT, called fast Fourier transform (FFT), is used, which reduces the calculation time from being proportional to n2 to being proportional to nlogn. Fourier and other transforms are also used in image compression, feature detection, and texture characterization.

A group of image operations often used in preprocessing stages of applications consists of functions that enhance or smooth an image, reduce image noise, or change contrast. These techniques include histogram equalization, where the image histogram is computed and then pixel intensities are adjusted to achieve uniform distribution of histogram frequencies, contrast modification such as the gamma function mentioned above, and image sharpening, for example using edge enhancement, or softening by a Gaussian filter.

Image conversion operations are used to change image format, range, precision, or color representation. They include conversion of type, such as integer to floating point; precision, for example, 8 bits to 1 bit; color representation, such as red-green-blue (RGB) to hue-saturation-value (HSV); or number of color channels, for example, RGB to gray scale.

8.2.2 Detection Methods

Detection methods are functions that locate parts of an image with desired characteristics. Interest operators seek point-like local features that are distinct in their neighborhood. Other types of detectors include edge, ridge, blob, and region detectors. A Hough transform is a technique to find more specific classes of shapes such as straight lines by a voting procedure (Ballard 1981; Duda and Hart 1972).

Corner Detectors Harris is a derivative-based corner detector (Harris and Stephens 1988). It is based on second moment matrix often used for feature detection and for describing local image structures. SUSAN detector based on efficient morphological operators (Smith and Brady 1997) can be used for corner and edge detection and also noise suppression. Harris and SUSAN detectors are invariant under translation and rotation. Harris extensions with higher levels of invariance are Harris-Laplace, a scale-invariant version, and Harris-Affine, an affine-invariant corner detector (Mikolajczyk and Schmid 2004).

Blob Detectors Hessian detector, like Harris, builds on the derivative information in the image. It has also extensions: scale invariant (Hessian-Laplace) and affine invariant (Hessian-Affine) (Mikolajczyk and Schmid 2004). Salient region detector is based on entropy of the intensity probability distribution (Kadir and Brady 2001).

Region Detectors Intensity-based region IBR detector starts from intensity extrema and explores the image in a radial way (Tuytelaars and Van Gool 2000, 2004) and stops on each ray in the place with significant change. MSER detector (Matas et al. 2004) extracts homogeneous intensity regions which are stable over a wide range of thresholds. Maximally stable extremal regions are extracted with a watershed-like segmentation algorithm.

Efficient Implementations Methods described above involve the computation of derivatives or more complex measures such as the second moment matrix. There are more effective detectors. Difference-of-Gaussian (DoG) detector extracts blobs in the image by approximating the Laplacian using multiple scale-space pyramids (Crowley and Parker 1984). SURF detector (Bay et al. 2006) makes use of integral images (Viola and Jones 2001) to efficiently compute a rough approximation of the Hessian matrix. FAST detector (Rosten and Drummond 2005, 2006) builds on the SUSAN detector evaluating only a limited number of individual pixel intensities using decision trees ID3 algorithm (Quinlan 1986).

Examples of Detection Methods in Membrane and Cell Images
Gebäck and Koumoutsakos (2009) describe a method for edge detection based on the discrete curvelet transform. They prove it useful for finding edges and elongated structures in images where the edges may not easily be detected using Canny detector (Canny 1986) and based on Gabor filter edge detector methods. Seyedhosseini et al. (2011) propose of a multi-scale representation of the context image with so-called radon-like features (Kumar et al. 2010) to learn a series of discriminative models for membrane detection in electron microscopy (EM) large area images. Ortiz De Solorzano et al. (2006) apply 3D generalized Hough transform and morphological reconstruction (Vincent 1993) for nuclei detection.

8.2.3 Segmentation Methods

Image segmentation is an operation which partitions an image into contiguous regions. The segmentation can be based on conditions which relate to the properties of the interior of the region, such as a similarity of intensity or texture between neighboring areas inside the region, differences between the region and its neighboring regions, characteristics and shape of the region's boundary, or some combination of these features. Often, these conditions are not explicitly stated but defined implicitly by the segmentation method. Many approaches to image segmentation have been proposed (Gonzalez and Woods 2017; Russ and Neal 2016), and new more advanced methods are added frequently. The following are some of the often used image segmentation techniques:

Intensity thresholding methods assign to each pixel p a label k if $T_{k-1} <= I(p) < T_k$, where $I(p)$ is the intensity of the pixel p and T_{k-1} and T_k are the lower and upper thresholds for the label k. Usually $T_0 = 0$ and $T_{kmax} = 255$. If $k$max = 2, then the result of this operation is a binary image. The threshold values can be set manually or automatically, for example, by finding the local minima of intensity histogram. The thresholds may be global, i.e., identical for every pixel or adjusted locally (Gonzalez and Woods 2017), for example, based on local histograms. The pixels having identical label may be grouped into contiguous regions, for example, using the connected component algorithm (Gonzalez and Woods 2017).

Region growing approach begins with selection of "seeds," consisting of single pixels or small segments, which form the initial regions. Iteratively, neighboring pixels or image patches are being added to the current regions if they satisfy a homogeneity criterion that determines that the added pixel or patch is sufficiently similar to the region. An example of such criterion is a thresholded absolute difference between the pixel value and an average of the region patch nearest to that pixel.

Region merging begins with an initially segmented image, for example, a segmentation where each pixel is a separate region. Region merging proceeds iteratively; in each step, pairs of adjacent similar regions are merged (Nock and Nielsen 2004). Variants of this method may use different strategies to determine the order of merging operations. Region merging may be applied to the result of region growing to reduce the number of regions and make adjacent regions more dissimilar.

Watershed methods (Barnes et al. 2014; Beucher and Meyer 2017) are mostly based on an analogy of flooding a topographic surface. A water source is set in each valley and the area is gradually flooded. The points where waters from different sources meet are collected into "watershed lines," which form the boundaries between segments. In such a basic formulation, the watershed method leads to oversegmentation and results with many small segments, due to image noise. The number of segments can be reduced by adding a condition that limits the water sources to a subset of valleys or follow the watershed method with region merging (Bleau and Leon 2000; Haris et al. 1998). Variants of this method use the image gradient (Gauch 1999) instead of the original image.

Clustering segmentation is based on a clustering operation (Jain and Dubes 1988), which groups pixels into clusters, using a predefined distance measure. Its aim is to obtain a grouping with a low average intra-cluster distance and high average inter-cluster distance. A clustering-based segmentation (Jain and Flynn 1996) can apply any general clustering method to image pixels, where a pixel-to-pixel distance is a function of weighted differences of intensity, location, and results of pixel neighborhood operators, such as texture features.

Active contour segmentation is based on a region model that accounts for the properties of regions, their boundaries or contours, and image edges. A parametric active contour, or a snake model (Caselles et al. 1993, 1997; Kass et al. 1988), defines an objective function represented as energy, $E_{snake} = E_{image} + E_{gradient} + E_{contour}$, where the components E_{image}, $E_{gradient}$, and $E_{contour}$ assess, respectively (with lower values being better), the accuracy of image approximation by the region model, the quality of the match between the region contours and edges within the image edges, and the smoothness of the contour. The segmentation is a result of minimizing E_{snake} functional. The underlying concept here is that each of these three terms attempts to shape the contour in a different way and the result is a balanced compromise.

A newer formulation of a geometric active contour model (Chan and Vese 2001) is also based on minimization of energy functional, but it does not include the edge component. The minimization problem is constructed using a level set method (Osher and Sethian 1988), which defines a function $F(x, y, t)$, related to the energy functional, where $x;y$ is a pixel location and t represents time, which evolves in time according to a specified partial differential equation. The contours formed by a set of $x;y$ positions where the energy functional is minimized and $F(x, y, t) = 0$ give the result of the segmentation.

In a *graph partitioning-based segmentation*, the image is represented as a graph, where each pixel is a graph vertex and edges are placed between vertices corresponding to adjacent edges. Image segmentation is formulated as a graph partitioning problem, where some edges are removed or "cut," segmenting the graph into mutually disjoint subgraphs (Felzenszwalb and Huttenlocher 2004; Shi and Malik 2000). The segmentation objective can be formulated as energy minimization or maximum flow, and many of such criteria can be mapped to the minimum cut problem for which efficient solutions exist (Boykov and Funka-Lea 2006; Boykov and Kolmogorov 2004).

In a *trainable segmentation* approach, the segmentation method learns from provided examples of segmented images, where the separated regions are annotated with a region index, for example, by a human expert. For each pixel and its neighborhood, treated as a sample for classification, a number of features are computed, for example, responses of linear and nonlinear filters, parameters of a locally fitted function such as a 2D polynomial, or parametric attributes of local image texture. These features are used as input vectors to a method that trains a pixel classifier on images with annotated regions; then the same features are used by a trained classifier to assign each pixel to one of the regions (Arganda-Carreras et al. 2017).

Examples of Segmentation Methods in Membrane and Cell Images

Dimopoulos and his research team (2014) propose pattern-based cell segmentation solution that can effectively and accurately segment densely packed cells.

For segmentation of biological neuron membranes, Dan Ciresan et al. (2012) use a special type of deep artificial neural network as a pixel classifier, where the input features are the image intensities in a square window centered at the pixel. They also provide a short survey on segmentation of neuronal membrane. Mikula et al. (2011) present partial differential equations (PDE) and level set methods for segmentation objects from 3D cell membrane images. They extract the inner cell boundaries, the surface of the organism, and the intercellular borders. Prodanov and Verstreken (2012) present multiple algorithms implemented in ImageJ for segmentation of cell structures. Wdowiak et al. (2015) use binary nuclei mask created by the classifier for the cell nuclei segmentation and intensity maps for membrane segmentation. Histace et al. (2014) use the approximate entropy embedded with geodesic active contour framework for application to membrane segmentation and present results obtained on confocal microscopy images. The CellECT tool (Delibaltov et al. 2016) provides functionalities based on watershed method for convenient segmentation and analysis of 3D+t membrane datasets. It can combine human interaction with automated algorithms. CellSegm (Hodneland et al. 2013) is another framework in a form of MATLAB toolbox for 3D cell segmentation. The algorithm aims at obtaining correct segmentation with minimum user interaction. Erik Meijering (2012) presents a brief description of classical cell segmentation methods. Boukari and Makrogiannis (2015) introduced nonlinear spatiotemporal diffusion-based motion detection for cell segmentation in time-lapse image sequences. In this framework, the linear diffusion model is equivalent to applying Gaussian filtering to the image. Authors applied Parzen kernel-based discontinuity detection (Parzen 1962) to produce stochastic multivalued edge map and watershed segmentation to detect moving cells. Another method (Frolkovič et al. 2007) for cell segmentation uses geometrical advection-diffusion equations. In Tscherepanow et al. (2008), bright-field images are segmented by means of an active contour with greedy approach (Williams and Shah 1992), morphology, and classification with 90% recognition rates. A new method that uses recursive balanced graph partitioning to segment foreground components using a fast and efficient binarization has been proposed and implemented using ITK for fast segmentation of 3D cell nuclei (Arz et al. 2017). A novel automated method able to segment closely juxtaposed or touching cell nuclei obtained from 3D microscopic images (Li et al. 2007) is composed of three steps: gradient diffusion procedure, gradient flow tracking and grouping, and local adaptive thresholding. Graphical model approach using inference in a Bayes network for segmentation of multicell images acquired by fluorescence microscopy (Chen et al. 2006) is a fast and accurate segmentation method. It is especially useful for segmenting a field containing cells that are touching each other, as is often the case with yeast images. Bengtsson et al. (2004) presented a review on cell image segmentation using multiple classical methods. Hong Gao (2013) presented an overview with newer segmentation methods for cell structures including snake, level set, Chan-Vese, and graph partitioning.

8.2.4 Pattern Classification and Machine Learning Methods

The role of pattern classification (or pattern recognition) methods is to assign data objects, such as sets of measurements or observations, to classes. The function that performs such assignments, or a classifier, is generally constructed automatically, based on provided examples of data objects, using methods of machine learning (Bishop 2006; Goodfellow et al. 2016; Hastie et al. 2001). A large group of classification and machine learning methods restrict the representation of a data object to a fixed size feature vector. Its elements, called features, are computed from a data object. For example, a shape of an image object described by the object's boundary could be represented by a vector of four features, such as area, perimeter, length, and the lengths of the major and minor axes of the enclosing ellipse. Some of the most popular classification and learning methods are:

Nearest neighbor classifier finds in a pattern database the object, which is most similar or nearest to the classified sample, and assigns the class of the found object to the sample.

Naive Bayes classifier uses a conditional probability model $P(f_1; f_2; ::::; f_n|C_i) = P(f_1|C_i)P(f_2|C_i); ::::;P(f_n|C_i)$, where features $f_1; f_2:::$ are conditionally independent given a class C_i. It assigns a sample to the class with maximal posterior probability computed using the Bayesian formula. The learning method estimates the conditional $P(f_j|C_i)$ and class $P(C_i)$ probabilities. This classifier gives poor results when the conditional independence cannot be assumed.

Decision tree classifier starts at the root of a decision tree, successively assigning a sample to lower branches, based on the result of a test executed at each node using a subset of the features, until reaching a leaf node. The decision tree is constructed in a learning phase by recursively splitting the dataset at each node, based on a statistic computed from a group of features selected for this node. Decision tree classifier is usually better than naive Bayes for data with dependencies.

Random forest classifier is a set or an ensemble of decision trees. Classification answers computed independently by the decision trees are combined by voting or another decision aggregation method.

Support vector machine (SVM) classifier divides the feature space using hyperplanes placed halfway between groups of class examples. Variants of SVM classifier can perform feature space separation using a nonlinear hypersurface instead of a hyperplane (Cortes and Vapnik 1995). Learning method for SVM efficiently solves a constrained optimization problem by quadratic programming.

Perceptron is also a linear classifier, but in contrast to SVM, it can be trained incrementally or online, adding new training samples one at a time.

Artificial neural network (ANN) is a multilayer network of perceptrons, which are modified by approximating their discontinuous threshold function with a differentiable function, which enables the use of a gradient descent-based optimization technique called backpropagation for training ANN.

Deep learning methods (Goodfellow et al. 2016) can be considered to be an extension and refinement of ANN. They use multilayer networks composed of different types of nonlinear processing elements. These networks are carefully structured and controlled. In addition to backpropagation, they might also use other learning methods, and they may be implemented using special hardware, for large processing tasks. Deep learning methods have achieved considerable successes in applications ranging from computer vision to speech recognition and natural language processing.

Examples of Classification and Machine Learning Methods in Membrane and Cell Images

A survey of the computer vision and machine learning methods for generating and categorizing phenotypic profiles of cells is presented in Grys et al. (2016). Macke et al. (2008) develop contour propagation model minimizing an energy function for finding the cell membranes. However, this active contour model can get stuck in local minima due to the complex intracellular structures and may find false boundaries (Mishchenko 2009). Vu and Manjunath (2008) proposed a graph cut framework that minimizes an energy defined over the image intensity and the intensity gradient field. But, the graph cut method might be misled by the complex intracellular structure of the EM images and requires the user to correct segmentation errors. Convolutional neural network proposed by Turaga et al. (2010) is capable for restoring membranes in EM images. Jurrus et al. (2010) proposed a framework to detect neuron membranes which integrates information from the original image together with contextual information by learning a series of artificial neural network (ANN). This makes the network much easier to train because the classifiers in the series are trained one at a time and in sequential order. Review of creation of accurate models of cell organization directly from images is proposed in Murphy (2016). Kasson et al. (2005) developed a novel learning-based method for the classification of plasma membrane protein localization data obtained via fluorescence microscopy and the differentiation of these data from intracellular artifacts.

8.2.5 Registration Methods

The purpose of medical image registration is to provide – at minimum – two images (one reference and one or more sensed) from different modalities that are spatially consistent (Brown 1992). Registration by executing specified transformations modifies the source image toward the reference one. General steps of image registration process between two images contain (Zitova and Flusser 2003):

1. *Feature detection.* Distinctive features like points, lines, line intersections, edges, corners, contours, and regions from both images are detected automatically or manually, and feature points are represented by their descriptors.

2. *Feature matching.* Corresponding types of features detected from sensed and reference images have been aligned using feature detector and similarity measures.
3. *Transform model estimation.* Estimation of the type and parameters of mapping function between a reference and target image.
4. *Image resampling and transformation.* Transformation of sensed image using estimated mapping function.

There are plenty of ways for classifications of image registration techniques (Sharma and Goyal 2013). Good survey on registration algorithms has been presented in (Sindhu Madhuri 2014). Survey on general nonrigid registration approaches is presented in Holden (2008).

Examples of Registration Methods in Membrane and Cell Images
Marco Tektonidis et al. (2015) present a nonrigid multi-frame registration approach for live cell fluorescence microscopy image data. Such an approach significantly improves the registration accuracy and is more robust to image noise and intensity scaling compared to standard pairwise registration. Registration is also useful during tracking process. It has been successfully implemented at pixel level (Hand et al. 2009), feature points (Matula et al. 2006), or whole cell level (Wilson and Theriot 2006). Würflinger et al. (2004) and Wilson and Theriot (2006) have used intensity-based image registration approach for cell nuclei and cell images. To compute 2D translation for the registration of intravital video microscopy images of rolling leukocytes, Gobic et al. used (2005) a correlation-based approach. Baheerathan et al. (1998) used the phase correlation method to determine rigid transformations and a landmark-based approach for computing affine transformations of serial sections of mouse liver cell nuclei. To compute 3D translation and rotation of cell nuclei to register labeled proteins, Rieger et al. (2004) defined the center of mass and the inertia tensor. Point-based solutions to compute 3D rotation and translation of live cells have been used by Matula et al. (2006) and Gerlich et al. (2001). Also a few approaches for nonrigid registration of cell nuclei images have been described. Some of them use splines and extracted point landmarks (Mattes et al. 2006) or the extension of the demon algorithm (Thirion 1998) using segmented images (Yang et al. 2008) or original images (Kim et al. 2007). Most of intensity-based approaches for nonrigid registration of dynamic cell nuclei images are based on the demon algorithm. Il-Han Kim et al. (2011) presented a novel intensity-based approach based on the Lucas-Kanade optic flow algorithm (Lucas et al. 1981) for temporal registration of 2D and 3D multichannel fluorescence cell microscopy images.

8.2.6 Tracking Methods

Cell migration and motility is an essential characteristic to understand the metastatic dissemination of cancer cells and various physiological processes like immune response or embryonic development. It is also a crucial component of biological research and drug discovery methods that support the identification of substances in living organisms affecting a phenotype of living cells. One of the most common ways of analyzing the cell motility is the use of tracking methods, which determine cell trajectories and significant changes of cell characteristics in live cell imaging data.

Many tracking methods and tools using robust techniques from image processing domain were developed during the last years and applied in various cell analysis procedures. There are several distinguishable approaches in cell tracking domain, and some of the most often used are described below.

Object association models use two separate steps to track the cells over time: the first one is cell detection in particular frames using segmentation methods, while the second task is association of cells between consecutive frames to obtain trajectories. Baker et al. (2014) use automated contour-based tracking for analysis of cell behavior in complex in vitro environments. They use segmentation to identify a cell as a first step and then cells in consecutive images using the Kilfoil linking system (Gao and Kilfoi 2009). They developed a new tracking tool for cell motility analysis in low-contrast images over long time scales, which is able to detect multicell interactions like cell divisions and merging events. Their powerful approach for fundamental understanding of the effect of cell-cell interactions combined with quantitative statistical metrics was expected to provide new insights into cancer biology, mechanobiology, and morphogenesis. Perner (2016) proposes a similarity-based tracking method for high-content screening in drug discovery and computational biology experiments, which can track the cells without an initialization of the algorithm parameters. The image is initially processed by thresholding segmentation and morphological filtering, the normal cells and mitotic cells are differentiated by use of texture classifier, and then similarities are determined between extracted cell bounding boxes areas based on average minimal pixel distances and used for tracking in successive time frames. Obtained paths are used to extract features that describe the motilities and velocities of cells to study the cell kinetics.

Deformable models are based on object detection step at the beginning usually obtained by explicit parametric contours or implicit level set segmentation methods and updating this model in consecutive frames to track this object over some period of time. Zimmer et al. (2002) propose a tracking method for quantitative analysis of cell dynamics from in vitro videomicroscopy data using parametric active contours model to track the low-contrast boundary deformations of moving cells in consecutive frames while requiring the initial segmentation in the first frame. Li et al. (2008) developed a fully automated multi-target tracking system for thousands of cells in time-lapse contrast microscopy. They use a fast topology-constrained geometric active contour tracker, adaptive motion filtering, and spatiotemporal trajectory optimization. Their method enables the automatic cell migration quantification and cell lineage map construction supporting the massive biological dataset analysis.

Learning-based models use classification methods learned from annotated training data for cell tracking enhancement. Lou and Hamprecht (2011) use learning to track a large quantity of cells in cell culture study and developmental biology, which allows to find the optimum parameters automatically from a training set for rich set of features. They introduce a generalization of object association models, which improves the expressiveness of the model and increases the number of parameters, and next use structured learning for automatic learning of the optimum parameters from a training set maximizing the profit from rich description. Sebag et al. (2015) use cell tracking for generic methodological workflow to analyze the single cell motility in high-throughput time-lapse screening data. They characterize cells by a set of 230 features including geometric, shape, and texture and then match cells in consecutive frames by determining the most likely instant temporal behavior. The search over all possible matches between cells is supported by distance thresholding and weightening learned by a support vector machine (SVM) using annotated trajectories. Captured trajectories are described by a set of 15 features representing partial stochasticity of cell migratory behavior. Additionally some global descriptors of the cell trajectory are calculated based on group of features like convex hull area, largest move on particular trajectory, and average track curvature. Such newly designed trajectory features and an original statistical procedure allow for identification of an ontology of cell motility patterns in an unsupervised manner with inferring the motion types from the data without use of any prior knowledge.

8.2.7 Visualization

Appropriate visualization methods play the important role in analysis of digital cell imaging data. They provide the biologists with very powerful and efficient tools for handling the complex and large data resources in an easy and intuitive way. Unfortunately there are not too many comprehensive guides collecting and summarizing available visualization methods in a consistent way because specific cell visualization methods are assigned to specific research groups working most often in the biomedical domain and in isolation from dominant data visualization fields like image processing. Among works addressing this gap, O'Donoghue et al. (2010) discuss the challenges of visualizing biological data and available visualization tools. Walter et al. (2010) present an overview of existing visualization methods from cells to organisms and tools with emphasizing their limitations and challenges. They (1) describe the benefits of digital representation of images in relation to its analog counterparts and explain most important aspects for visualization of high-dimensional image data; (2) provide an overview of current image file formats, treatment of the time dimension, and use of image processing methods in visualization like segmentation and registration; and (3) discuss current implementation issues.

Pretorius et al. (2016) provide a structured qualitative analysis of visualization methods describing the domain, data, and abstract pipeline of tasks in visualization techniques from visualization theory to its practical applications. This allows the biologists, visualization developers, and visualization researchers to evaluate existing visualization methods and learn the current research results to identify the current shortcomings and prioritize their efforts to overcome them.

Analyzing the particular visualization methods, several classes can be defined (Pretorius et al. 2016) depending on what kind of visualized aspects are emphasized including positions of objects, different data properties, temporal changes, aggregate behavior, descendant relationships, or abstract data derived from the postprocessing phase.

Spatial embedding methods allow the users to interpret their data in intuitive way by representing the spatial locations where activity occurred. They support the analysis of cell growth, shape, movement, and reproduction from acquired images and help to understand their impact on cell fate, formation of tissue, and organs. Gerlich et al. (2003) present the concepts of automated multidimensional image data analysis for live cell microscopy and the dynamics of cell nuclear subcompartments. Fangerau et al. (2012) offer a method for exploring 2D projections of 3D cell movement video, which allows the biologists to analyze similarities and differences of division patterns and cell migration over the entire organism.

Space-time cube methods allow to present the results of image processing algorithms in specific cases and to compare them with other types of analysis. Meijering et al. (2006) summarize the problems and limitations of the tracking methods in biological molecular imaging, while Molenaar et al. (2003) perform a quantitative motion analysis of time-lapse image sequences with the use of diffusion filtering, adaptive thresholding, and tracking algorithm to present and confirm the results generated by method using peptide nucleic acid (PNA) probes for investigation of telomeres in living cells.

Temporal plot methods are used to analyze the temporal properties of cellular behavior, such as proposed by Scherf et al. (2012), who developed an automated framework for image analysis and colony tracking to obtain a quantification of structural properties of cell colonies evolving in space and time.

Aggregate visualization methods allow to analyze cellular behavior at higher levels than single cell including cell responses to perturbations or cell colony behavior. Pretorius et al. (2015) address an interactive visualization of spatiotemporal behavior of cell lineages including variation from average behavior and presence or absence of symmetry and synchrony, which describes the lineage branching structure and temporal alignment of cellular events. Duffy et al. (2015) use glyphs to encode the numerical measurements and to summarize spatiotemporal motion characteristics by use of static visual representations in video visualization task for computer-aided semen analysis.

Dimension reduction methods are based on analyzing the statistics of live cell imaging results. Hamilton et al. (2009) introduce a method of statistical testing and deliver suitable software for testing the differences in subcellular imaging. Strobelt et al. (2012) propose a visualization tool for examining the biochemical processes, which allows for the projection of one or a few molecules of interest.

Lineage diagram methods analyze temporal development of cells, cell division and death, as well as the relation of these aspects with each other. Glauche et al. (2009) use statistical analysis of cellular genealogies addressing degree and symmetry of cellular expansion and occurrence and correlation of characteristic events such as cell death to reconstruct the lineage fate decisions, which is essential to understand the mechanisms of lineage commitment.

The above described classification does not exhaust all cell visualization use cases, which are often used in a complementary way with other visualization phenomena like protein colocalization. Hou et al. (2015) propose a spatiotemporal measurement tools for dynamic cell analysis with application in cancer metastasis using graph cut (max-flow and min-cut), heatmaps, and segmentation methods. They use a technique to classify cell velocities into different states (protrusion, quiescence, and retraction) and visualize their occurrence and localization on videos of migrating cells. Additionally they propose a semiautomatic tool for selecting regions of interest (ROIs) based on correlation maps and plot these ROIs onto the original cell migration video, which helps the biologists to observe important migration regions in a user-friendly way. Finally their visualization method includes the correlation between cell membrane dynamics and subcellular protein colocalization, which allows to investigate the relationships between short-term cell membrane dynamics, protein colocalization, and long-term cell invasion behavior.

8.2.8 Software for Biological Image Analysis

This part describes selected open-source software and libraries for specialized biological image analysis.

ImageJ (Schindelin et al. 2015; Schneider et al. 2012) from the National Institutes of Health (NIH) is one of the best known open-source software packages for biological image analysis. It is dedicated mainly for analysis of individual images. *Fiji* (Schindelin et al. 2012) is based on stable ImageJ distribution with many bundled plug-ins. Users can extend ImageJ and Fiji by developing their own extensions (Broeke et al. 2015) in ImageJ Macro, JavaScript, BeanShell (Niemeyer 2008), Jython (Jython 2008), JRuby (JRuby 2008), Clojure, Groovy (Groovy 2008), and Scala (Odersky et al. 2011). A good starting point with resources for programmers can be found in ImageJ (2017).

There are several packages using ImageJ like a core element for specific biological image processing and more high-throughput work. Icy (De Chaumont et al. 2012) is a platform that allows to visualize, annotate, and quantify bioimaging 2D and 3D data and native integration with ImageJ. In 2006, Anne E Carpenter et al. (2006) presented *CellProfiler*, an image analysis open-source software for identifying and quantifying standard and complex morphological assays and tasks like analysis of subcellular patterns and cell or organelle shape. It is dedicated to 2D cell images and is a great alternative to commercial software. Using features extracted by CellProfiler, one can use *CellProfiler Analyst* (Jones et al. 2008, 2009) for exploration and mining of cell image data, generated in ever-increasing amounts in high-content screens. CellProfiler Analyst 1.0 has been written in *Java* and allows to recognize a single phenotype in individual cell images. Version 2.0 described in Dao et al. (2016) has been rewritten in *Python* and uses *scikit-learn* machine learning library (Pedregosa et al. 2011) for classification of multiple phenotypes using popular models like AdaBoost, random forest, and SVM. Next source software package for analyzing, processing, and visualizing multidimensional microscopy images *BioImageXD* (Kankaanpää et al. 2012) is written in *Python* and *C++*, using *wxPython* for the GUI, and leverages the power of the Visualization Toolkit (VTK) (Schreibmann 2015) for multidimensional image processing and 3D rendering. BioImageXD also uses the most popular medical image analysis toolkit from NIH, Insight Segmentation and Registration Toolkit (*ITK*) (Johnson et al. 2017a, b; Yoo et al. 2002), for segmentation, registration, and other image processing tasks. Another open-source biological image classification software contains *Ilastik* (Sommer et al. 2011), *CellCognition* (Held et al. 2010), a command-line utility *Wndchrm* (Shamir et al. 2008), and *imageHTS* (Pau et al. 2013) – an *R* (Team 2016) package dedicated to the analysis of high-throughput microscopy-based screens. Ilastik is a tool for pixel-based classification of 2-, 3-, and 4D images where the user first trains a classifier by identifying areas of images that fall into one of several classes, such as cell body, nucleus, background, or membrane, and next applies it to those and similar images to identify areas in those images corresponding to the trained classes. CellCognition framework is dedicated to the automatic analysis of live cell imaging data in the context of high-content screening (HCS). The imageHTS provides a modular and extensible tool to segment cells, extract quantitative cell features, and predict cell types. The framework *STOCHASTIC* was created for dynamic image-based phenotypic profiling and is presented in Gordonov et al. (2016). Bio-Image Semantic Query User Environment (*BisQue*) (Kvilekval et al. 2010) – one of the few cloud-based integrated systems for storing, visualizing, organizing, and analyzing biological images in the form of web-based platform – integrates other mentioned software tools. A second one with client-server architecture is *OMERO* (Swedlow 2007) – a research data management platform for biology image files with flexible API for third-party data analysis tools.

Erlend Hodneland et al. (2013) published *CellSegm* – a MATLAB software toolbox for automated whole cell segmentation of images showing surface stained cells, acquired by fluorescence microscopy. The authors claim that it can be also used to detect various types of surface stained cells in 3D. The MATLAB-based software tool *OMAL* (Gudla et al. 2008) has been developed for the manual and automated segmentation of cells and cell nuclei at Frederick National Lab. *CellTracker* (Piccinini et al. 2015) is an image processing software to perform automated, semi-automated, and manual cell migration detection on phase-contrast, DIC, and fluorescent images.

CellECT is an interactive cell analysis tool for 3D time sequence microscopy dataset analysis, created by Diana L. Delibaltov et al. (Delibaltov et al. 2016). The main segmentation tool is watershed based – it interactively allows the user to add, remove, or modify discovered cell segments.

An API from *ImageJ Ops* (Rueden et al. 2017) enables programmers to code reusable image processing algorithms in the Ops framework available in ImageJ, CellProfiler, KNIME, OMERO, and Alida software projects. Reviews of various tools for biological image processing in various contexts were presented in Eliceiri et al. (2012) and Barry et al. (2015) and their usability in Carpenter et al. (2012). A great book titled *Bioimage Data Analysis* which is suited for biologists and providing directions how to analyze biomedical images in examples has been published by Kota Miura (2016) as an open-access textbook.

References

Arganda-Carreras I, Kaynig V, Rueden C, Eliceiri KW, Schindelin J, Cardona A, Sebastian Seung H (2017) Trainable weak segmentation: a machine learning tool for microscopy pixel classification. Bioinformatics. Oxford, doi: 10.1093/bioinformatics/btx180

Arz J, Sanders P, Stegmaier J, Mikut R (2017) 3d cell nuclei segmentation with balanced graph partitioning. arXiv preprint arXiv:1702.05413

Baheerathan S, Albregtsen F, Danielsen HE (1998) Registration of serial sections of mouse liver cell nuclei. J Microsc 192(1):37–53

Baker RM, Brasch ME, Lisa Manning M, Henderson JH (2014) Automated, contour-based tracking and analysis of cell behaviour over long time scales in environments of varying complexity and cell density. J R Soc Interface 11(97):20140386

Ballard DH (1981) Generalizing the hough transform to detect arbitrary shapes. Pattern Recogn 13(2):111–122

Barnes R, Lehman C, Mulla D (2014) Priority-flood: an optimal depression-filling and watershed-labeling algorithm for digital elevation models. Comput Geosci 62:117–127

Barry DJ, Durkin CH, Abella JV, Way M (2015) Open source software for quantification of cell migration, protrusions, and fluorescence intensities. J Cell Biol. doi:10.1083/jcb.201501081

Bay H, Tuytelaars T, Van Gool L (2006) Surf: speeded up robust features. In: Computer vision – ECCV 2006. Springer, Berlin, Heidelberg pp 404–417

Bengtsson E, Wahlby C, Lindblad J (2004) Robust cell image segmentation methods. Pattern Recognit Image Anal 14(2):157–167

Beucher S, Meyer F (2017) The morphological approach to segmentation: the watershed transformation. In: Mathematical morphology in image processing, 1st edn. Marcel Dekker Inc, New York, pp 433–481

Bishop CM (2006) Pattern recognition. Mach Learn 128:1–58

Bleau A, Leon LJ (2000) Watershed-based segmentation and region merging. Comput Vis Image Underst 77(3):317–370

Boukari F, Makrogiannis S (2015) Spatio-temporal diffusion-based dynamic cell segmentation. In: Bioinformatics and Biomedicine (BIBM), 2015 IEEE International Conference on, Washington, DC, USA, pp 317–324. IEEE

Boykov Y, Funka-Lea G (2006) Graph cuts and efficient nd image segmentation. Int J Comput Vis 70(2):109–131

Boykov Y, Kolmogorov V (2004) An experimental comparison of min-cut/max-flow algorithms for energy minimization in vision. IEEE Trans Pattern Anal Mach Intell 26(9):1124–1137

Broeke J, Pérez JMM, Pascau J (2015) Image processing with ImageJ. Packt Publishing Ltd, Birmingham

Brown LG (1992) A survey of image registration techniques. ACM Comput Surv 24(4):325–376

Canny J (1986) A computational approach to edge detection. IEEE Trans Pattern Anal Mach Intell PAMI-8(6):679–698

Carpenter AE, Jones TR, Lamprecht MR, Clarke C, Kang IH, Friman O, Guertin DA, Chang JH, Lindquist RA, Moffat J et al (2006) Cellprofiler: image analysis software for identifying and quantifying cell phenotypes. Genome Biol 7(10):R100

Carpenter AE, Kamentsky L, Eliceiri KW (2012) A call for bioimaging software usability. Nat Methods 9(7):666

Caselles V, Catté F, Coll T, Dibos F (1993) A geometric model for active contours in image processing. Numer Math 66(1):1–31

Caselles V, Kimmel R, Sapiro G (1997) Geodesic active contours. Int J Comput Vis 22(1):61–79

Chan TF, Vese LA (2001) Active contours without edges. IEEE Trans Image Process 10(2):266–277

Chen S-C, Zhao T, Gordon GJ, Murphy RF (2006) A novel graphical model approach to segmenting cell images. In: Computational Intelligence and Bioinformatics and Computational Biology, 2006. CIBCB'06. 2006 IEEE Symposium on, Toronto, Ont., Canada, pp 1–8. IEEE

Ciresan D, Giusti A, Gambardella LM, Schmidhuber J (2012) Deep neural networks segment neuronal membranes in electron microscopy images. In: Advances in neural information processing systems, pp 2843–2851

Cortes C, Vapnik V (1995) Support-vector networks. Mach Learn 20(3):273–297

Crowley JL, Parker AC (1984) A representation for shape based on peaks and ridges in the difference of low-pass transform. IEEE Trans Pattern Anal Mach Intell PAMI-6(2):156–170

Dao D, Fraser AN, Hung J, Ljosa V, Singh S, Carpenter AE (2016) Cellprofiler analyst: interactive data exploration, analysis and classification of large biological image sets. Bioinformatics 32(20):3210–3212

De Chaumont F, Dallongeville S, Chenouard N, Hervé N, Pop S, Provoost T, Meas-Yedid V, Pankajakshan P, Lecomte T, Le Montagner Y et al (2012) Icy: an open bioimage informatics platform for extended reproducible research. Nat Methods 9(7):690–696

Delibaltov DL, Gaur U, Kim J, Kourakis M, Newman-Smith E, Smith W, Belteton SA, Szymanski DB, Manjunath BS (2016) Cellect: cell evolution capturing tool. BMC Bioinformatics 17(1):88

Dimopoulos S, Mayer CE, Rudolf F, Stelling J (2014) Accurate cell segmentation in microscopy images using membrane patterns. Bioinformatics 30(18):2644–2651

Duda RO, Hart PE (1972) Use of the hough transformation to detect lines and curves in pictures. Commun ACM 15(1):11–15

Duffy B, Thiyagalingam J, Walton S, Smith DJ, Trefethen A, Kirkman-Brown JC, Gaffney EA, Chen M (2015) Glyph-based video visualization for semen analysis. IEEE Trans Vis Comput Graph 21(8):980–993

Eliceiri KW, Berthold MR, Goldberg IG, Ibáñez L, Manjunath BS, Martone ME, Murphy RF, Peng H, Plant AL, Roysam B et al (2012) Biological imaging software tools. Nat Methods 9(7):697–710

Fangerau J, Höckendorf B, Wittbrodt J, Leitte H (2012) Similarity analysis of cell movements in video microscopy. In: Biological Data Visualization (BioVis), 2012 IEEE Symposium on, Seattle, WA, USA, pp 69–76. IEEE

Felzenszwalb PF, Huttenlocher DP (2004) Efficient graph-based image segmentation. Int J Comput Vis 59:167

Frolkovič P, Mikula K, Peyriéras N, Sarti A (2007) Counting number of cells and cell segmentation using advection-diffusion equations. Kybernetika 43(6):817–829

Gao H (2013) Segmentation of cell structures in fluorescence confocal microscopy images. PhD thesis, University of Central Lancashire

Gao Y, Kilfoi ML (2009) Accurate detection and complete tracking of large populations of features in three dimensions. Opt Express 17(6):4685–4704

Gauch JM (1999) Image segmentation and analysis via multiscale gradient watershed hierarchies. IEEE Trans Image Process 8(1):69–79

Gebäck T, Koumoutsakos P (2009) Edge detection in microscopy images using curvelets. BMC Bioinformatics 10(1):75

Gerlich D, Beaudouin J, Gebhard M, Ellenberg J, Eils R (2001) Four-dimensional imaging and quantitative reconstruction to analyse complex spatiotemporal processes in live cells. Nat Cell Biol 3(9):852–855

Gerlich D, Mattes J, Eils R (2003) Quantitative motion analysis and visualization of cellular structures. Methods 29(1):3–13

Glasbey CA, Horgan GW (1995) Image analysis for the biological sciences. Wiley, Chichester/New York

Glauche I, Lorenz R, Hasenclever D, Roeder I (2009) A novel view on stem cell development: analysing the shape of cellular genealogies. Cell Prolif 42(2):248–263

Gonzalez RC, Woods RE (2017) Digital image processing, 4th edn. ISBN: 9780133356724. Pearson, 1184 pp

Goobic AP, Tang J, Acton ST (2005) Image stabilization and registration for tracking cells in the microvasculature. IEEE Trans Biomed Eng 52(2):287–299

Goodfellow I, Bengio Y, Courville A (2016) Deep learning. MIT Press, Cambridge, MA

Gordonov S, Hwang MK, Wells A, Gertler FB, Lauffenburger DA, Bathe M (2016) Time series modeling of live-cell shape dynamics for image-based phenotypic profiling. Integr Biol 8(1):73–90

Groovy (2008) Groovy an agile dynamic language for the java platform. Technical report, Technical report. http://groovy.codehaus.org/. Accessed 2008.

Grys BT, Lo DS, Sahin N, Kraus OZ, Morris Q, Boone C, Andrews BJ (2016) Machine learning and computer vision approaches for phenotypic profiling. J Cell Biol. doi:10.1083/jcb.201610026

Gudla PR, Nandy K, Collins J, Meaburn KJ, Misteli T, Lockett SJ (2008) A high-throughput system for segmenting nuclei using multiscale techniques. Cytometry A 73(5):451–466

Haidekker MA (2011) Advanced biomedical image analysis. Wiley, Hoboken

Hamilton NA, Wang JTH, Kerr MC, Teasdale RD (2009) Statistical and visual differentiation of subcellular imaging. BMC Bioinformatics 10(1):94

Hand AJ, Sun T, Barber DC, Hose DR, MacNeil S (2009) Automated tracking of migrating cells in phase-contrast video microscopy sequences using image registration. J Microsc 234(1):62–79

Haris K, Efstratiadis SN, Maglaveras N, Katsaggelos AK (1998) Hybrid image segmentation using watersheds and fast region merging. IEEE Trans Image Process 7(12):1684–1699

Harris C, Stephens M (1988) A combined corner and edge detector, In Proc. of Fourth Alvey Vision Conference, BMVA, University of Manchester, pp 147–151

Hastie T, Tibshirani R, Friedman J (2001) The elements of statistical learning. Vol. 1. New York: Springer series in statistics

Held M, Schmitz MHA, Fischer B, Walter T, Neumann B, Olma MH, Peter M, Ellenberg J, Gerlich DW (2010) Cellcognition: time-resolved phenotype annotation in high-throughput live cell imaging. Nat Methods 7(9):747–754

Histace A, Bonnefoye E, Garrido L, Matuszewski B, Murphy M (2014) Active contour segmentation based on approximate entropy: application to cell membrane segmentation in confocal microscopy. In Proceedings of the International Conference on Bio-inspired Systems and Signal Processing - Volume 1: BIOSIGNALS, (BIOSTEC 2014) ISBN 978-989-758-011-6, pp 270-277. doi:10.5220/0004903002700277

Hodneland E, Kögel T, Frei DM, Gerdes H-H, Lundervold A (2013) Cellsegm-a matlab toolbox for high-throughput 3d cell segmentation. Source Code Biol Med 8(1):16

Holden M (2008) A review of geometric transformations for nonrigid body registration. IEEE Trans Med Imaging 27(1):111–128

Hou Y, Cooper L, ScottWilkinson AM, Brat D (2015) Spatiotemporal visualization of cell membrane dynamics and protein colocalization reveals correlation between membrane dynamics and metastatic invasion. In: Proceedings of the 6th ACM Conference on Bioinformatics, Computational Biology and Health Informatics, Atlanta, Georgia, USA, pp 476–477. ACM

ImageJ (2017) Developer resources: macros, scripting, tutorials, api documentation. https://imagej.nih.gov/ij/developer/index.html

Jain AK, Dubes RC (1988) Algorithms for clustering data. Prentice Hall, Englewood Cliffs

Jain AK, Flynn PJ (1996) Image segmentation using clustering. IEEE Press, Piscataway

Jan J (2006) Medical image processing, reconstruction and restoration. Taylor & Francis/CRC Press, Boca Raton

Johnson HJ, McCormick MM, Ibanez L (2017a) The ITK software guide book 1: introduction and development guidelines-volume 1. Kitware, Inc. https://itk.org/ItkSoftwareGuide

Johnson HJ, McCormick MM, Ibanez L (2017b) The ITK software guide book 2: design and functionality-volume 2. Kitware, Inc. https://sourceforge.net/projects/itk/files/itk/4.11/InsightSoftwareGuide-Book2-7011.0.pdf/download

Jones TR, Kang IH, Wheeler DB, Lindquist RA, Papallo A, Sabatini DM, Golland P, Carpenter AE (2008) Cellprofiler analyst: data exploration and analysis software for complex image-based screens. BMC Bioinformatics 9(1):482

Jones TR, Carpenter AE, Lamprecht MR, Moffat J, Silver SJ, Grenier JK, Castoreno AB, Eggert US, Root DE, Golland P et al (2009) Scoring diverse cellular morphologies in image-based screens with iterative feedback and machine learning. Proc Natl Acad Sci 106(6):1826–1831

JRuby (2008) Jruby a java powered ruby implementation. Technical report, Technical report (2003–2011). http://jruby.org/. Accessed June 2011

Jurrus E, Paiva ARC, Watanabe S, Anderson JR, Jones BW, Whitaker RT, Jorgensen EM, Marc RE, Tasdizen T (2010) Detection of neuron membranes in electron microscopy images using a serial neural network architecture. Med Image Anal 14(6):770–783

Jython (2008) Jython implementation of the high-level, dynamic, object-oriented language python written in 100% pure java. Technical report, Technical report (1997–2011). http://www.jython.org/. Accessed June 2011

Kadir T, Brady M (2001) Scale, saliency and image description. Int J Comput Vis 45(2):83–105

Kankaanpää P, Paavolainen L, Tiitta S, Karjalainen M, Päivärinne J, Nieminen J, Marjomäki V, Heino J, White DJ (2012) Bioimagexd: an open, general-purpose and high-throughput image-processing platform. Nat Methods 9(7):683–689

Kass M, Witkin A, Terzopoulos D (1988) Snakes: active contour models. Int J Comput Vis 1(4):321–331

Kasson PM, Huppa JB, Davis MM, Brunger AT (2005) A hybrid machine-learning approach for segmentation of protein localization data. Bioinformatics 21(19):3778–3786

Kim I, Yang S, Le Baccon P, Heard E, Chen Y-C, Spector D, Kappel C, Eils R, Rohr K (2007) Non-rigid temporal registration of 2d and 3d multichannel microscopy image sequences of human cells. In: Biomedical Imaging: From Nano to Macro, 2007. ISBI 2007. 4th IEEE International Symposium on, Arlington, VA, USA, pp 1328–1331. IEEE

Kim I-H, Chen Y-CM, Spector DL, Eils R, Rohr K (2011) Nonrigid registration of 2-d and 3-d dynamic cell nuclei images for improved classi fication of subcellular particle motion. IEEE Trans Image Process 20(4):1011–1022

Kumar R, Vazquez-Reina A, Pfister H (2010) Radon-like features and their application to connectomics. In: 2010 IEEE Computer Society Conference on Computer Vision and Pattern Recognition – Workshops, San Francisco, CA, USA

Kvilekval K, Fedorov D, Obara B, Singh A, Manjunath BS (2010) Bisque: a platform for bioimage analysis and management. Bioinformatics 26(4):544–552

Li G, Liu T, Tarokh A, Nie J, Guo L, Mara A, Holley S, Wong STC (2007) 3d cell nuclei segmentation based on gradient flow tracking. BMC Cell Biol 8(1):40

Li K, Miller ED, Chen M, Kanade T, Weiss LE, Campbell PG (2008) Cell population tracking and lineage construction with spatiotemporal context. Med Image Anal 12(5):546–566

Lou X, Hamprecht FA (2011) Structured learning for cell tracking. In: Advances in neural information processing systems, pp 1296–1304

Lucas BD, Kanade T et al (1981) An iterative image registration technique with an application to stereo vision. In Proceedings of the 7th international joint conference on Artificial intelligence - Volume 2 (IJCAI'81), Vol. 2. Morgan Kaufmann Publishers Inc., San Francisco, CA, USA, 674-679.

Macke JH, Maack N, Gupta R, Denk W, Schölkopf B, Borst A (2008) Contour-propagation algorithms for semi-automated reconstruction of neural processes. J Neurosci Methods 167(2):349–357

Matas J, Chum O, Urban M, Pajdla T (2004) Robust wide-baseline stereo from maximally stable extremal regions. Image Vis Comput 22(10):761–767

Mattes J, Nawroth J, Boukamp P, Eils R, Greulich-Bode KM (2006) Analyzing motion and deformation of the cell nucleus for studying colocalizations of nuclear structures. In: Biomedical Imaging: Nano to Macro, 2006. 3rd IEEE International Symposium on, Arlington, VA, USA, pp 1044–1047. IEEE

Matula P, Kozubek M, Dvorak V (2006) Fast point-based 3-d alignment of live cells. IEEE Trans Image Process 15(8):2388–2396

Meijering E (2012) Cell segmentation: 50 years down the road [life sciences]. IEEE Signal Process Mag 29(5):140–145

Meijering E, Smal I, Danuser G (2006) Tracking in molecular bioimaging. IEEE Signal Process Mag 23(3):46–53

Mikolajczyk K, Schmid C (2004) Scale & affine invariant interest point detectors. Int J Comput Vis 60(1):63–86

Mikula K, Peyriéras N, Remešíková M, Stašová O (2011) Segmentation of 3d cell membrane images by pde methods and its applications. Comput Biol Med 41(6):326–339

Mishchenko Y (2009) Automation of 3d reconstruction of neural tissue from large volume of conventional serial section transmission electron micrographs. J Neurosci Methods 176(2):276–289

Molenaar C, Wiesmeijer K, Verwoerd NP, Khazen S, Eils R, Tanke HJ, Dirks RW (2003) Visualizing telomere dynamics in living mammalian cells using pna probes. EMBO J 22(24):6631–6641

Mura K (2016) Bioimage data analysis, free textbook. Wiley. http://www.imaging-git.com/olympus-website-bioimage-data-analysis

Murphy RF (2016) Building cell models and simulations from microscope images. Methods 96:33–39

Niemeyer P. (2008) Beanshell: lightweight scripting for java. Homepage: http://www.beanshell.org

Nock R, Nielsen F (2004) Statistical region merging. IEEE Trans Pattern Anal Mach Intell 26(11):1452–1458

O'Donoghue SI, Gavin A-C, Gehlenborg N, Goodsell DS, Hériché J-K, Nielsen CB, North C, Olson AJ, Procter JB, Shattuck DW et al (2010) Visualizing biological dataâ˘A˘ Tnow and in the future. Nat Methods 7:S2–S4

Odersky M, Spoon L, Venners B (2011) Programming in scala: a comprehensive step-by-step guide, 2nd edn. Artima Incorporation, Walnut Creek

Ortiz De Solorzano C, Malladi R, Lelievre SA, Lockett SJ (2006) Segmentation on nuclei and cells using membrane related protein markers. J Microsc 222(1):67

Osher S, Sethian JA (1988) Fronts propagating with curvature-dependent speed: algorithms based on hamilton-jacobi formulations. J Comput Phys 79(1):12–49

Parzen E (1962) On estimation of a probability density function and mode. Ann Math Stat 33(3):1065–1076

Pau G, Zhang X, Boutros M, Huber W (2013) imagehts: analysis of high-throughput microscopy-based screens. R package version 1.0

Pedregosa F, Varoquaux G, Gramfort A, Michel V, Thirion B, Grisel O, Blondel M, Prettenhofer P, Weiss R, Dubourg V et al (2011) Scikit-learn: machine learning in python. J Mach Learn Res 12(Oct):2825–2830

Perner P (2016) Microscopic imaging of living cells, automatic tracking and the description of the kinetics of the cells for image- mining. J Transl Sci 3. doi: 10.15761/JTS.1000168

Piccinini F, Kiss A, Horvath P (2015) Celltracker (not only) for dummies. Bioinformatics. 2016 Mar 15;32(6):955–7

Pratt WK (2007) Digital image processing. Wiley-Interscience

Pretorius AJ, Khan IA, Errington RJ (2015) Cell lineage visualisation. In: Computer Graphics Forum, vol 34. Wiley Online Library, pp 21–30

Pretorius AJ, Khan IA, Errington RJ (2016) A survey of visualization for live cell imaging. In: Computer Graphics Forum. Wiley Online Library

Prodanov D, Verstreken K (2012) Automated segmentation and morphometry of cell and tissue structures. Selected algorithms in ImageJ. INTECH Open Access Publisher

Quinlan JR (1986) Induction of decision trees. Mach Learn 1(1):81–106

R Core Team (2016) A language and environment for statistical computing. r foundation for statistical computing, 2015, Vienna

Rieger B, Molenaar C, Dirks RW, Van Vliet LJ (2004) Alignment of the cell nucleus from labeled proteins only for 4d in vivo imaging. Microsc Res Tech 64(2):142–150

Roeder AH, Cunha A, Burl MC, Meyerowitz EM (2012) A computational image analysis glossary for biologists. Dev 139(17):3071–80. doi: 10.1242/dev.076414

Rosten E, Drummond T (2005) Fusing points and lines for high performance tracking. In: Tenth IEEE International Conference on Computer Vision (ICCV'05), vol 1

Rosten E, Drummond T (2006) Machine learning for high-speed corner detection. In: Computer Vision âĂŞ ECCV 2006, pp 430–443

Rueden, Horn, Schindelin, Northan, Berthold, Eliceiri (2017) http://imagej.net/Ops

Russ JC, Neal FB (2016) The image processing handbook, 7th edn., crc press, boca raton, fl, 2015, 1053 pp. isbn: 978-1498740265. Microsc Microanal 22(03):733

Scherf N, Herberg M, Thierbach K, Zerjatke T, Kalkan T, Humphreys P, Smith A, Glauche I, Roeder I (2012) Imaging, quantification and visualization of spatio-temporal patterning in mesc colonies under different culture conditions. Bioinformatics 28(18):i556–i561

Schindelin J, Arganda-Carreras I, Frise E, Kaynig V, Longair M, Pietzsch T, Preibisch S, Rueden C, Saalfeld S, Schmid B et al (2012) Fiji: an open-source platform for biological-image analysis. Nat Methods 9(7):676–682

Schindelin J, Rueden CT, Hiner MC, Eliceiri KW (2015) The imagej ecosystem: an open platform for biomedical image analysis. Mol Reprod Dev 82(7–8):518–529

Schneider CA, Rasband WS, Eliceiri KW (2012) Nih image to imagej: 25 years of image analysis. Nat Methods 9(7):671

Schreibmann E (2015) We-d-201-03: Itk and vtk-the standard libraries. Med Phys 42(6):3672–3672

Sebag AS, Plancade S, Raulet-Tomkiewicz C, Barouki R, Vert J-P, Walter T (2015) A generic methodological framework for studying single cell motility in high-throughput time-lapse data. Bioinformatics 31(12):i320–i328

Seyedhosseini M, Kumar R, Jurrus E, Giuly R, Ellisman M, Pfister H, Tasdizen T (2011) Detection of Neuron Membranes in Electron Microscopy Images Using Multi-scale Context and Radon-Like Features. MICCAI 2011, Part I, LNCS, vol 6891, pp 670–677, Springer, Berlin, Heidelberg

Shamir L, Orlov N, Mark Eckley D, Macura T, Johnston J, Goldberg IG (2008) Wndchrm–an open source utility for biological image analysis. Source Code Biol Med 3(1):13

Sharma K, Goyal A (2013) Classification based survey of image registration methods. In: Computing, Communications and Networking Technologies (ICCCNT), 2013 Fourth International Conference on, Tiruchengode, India, pp 1–7. IEEE

Shi J, Malik J (2000) Normalized cuts and image segmentation. IEEE Trans Pattern Anal Mach Intell 22(8):888–905

Sindhu Madhuri G (2014) Classification of image registration techniques and algorithms in digital image processing–a research survey. Int J Comput Trends Technol (IJCTT) 15:78–82

Smith SM, Brady JM (1997) Susan – a new approach to low level image processing. Int J Comput Vis 23(1):45–78

Sommer C, Straehle C, Koethe U, Hamprecht FA (2011) Ilastik: interactive learning and segmentation toolkit. In: Biomedical Imaging: From Nano to Macro, 2011 IEEE International Symposium on, Chicago, IL, USA, pp 230–233. IEEE

Sonka M, Fitzpatrick JM (2009) Handbook of medical imaging: volume 2. Medical image processing and analysis, 1st edn. SPIE Press

Strobelt H, Bertini E, Braun J, Deussen O, Groth U, Mayer TU, Merhof D (2012) Hitsee knime: a visualization tool for hit selection and analysis in high-throughput screening experiments for the knime platform. BMC Bioinformatics 13(8):S4

Swedlow J (2007) The open microscopy environment: a collaborative data modeling and software development project for biological imageinformatics. In: Imaging cellular and molecular biological functions, Springer Berlin Heidelberg, pp 71–92

Tektonidis M, Kim I-H, Chen Y-CM, Eils R, Spector DL, Rohr K (2015) Non-rigid multi-frame registration of cell nuclei in live cell fluorescence microscopy image data. Med Image Anal 19(1):1–14

Thirion J-P (1998) Image matching as a diffusion process: an analogy with maxwell's demons. Med Image Anal 2(3):243–260

Tscherepanow M, Zöllner F, Hillebrand M, Kummert F (2008) Automatic segmentation of unstained living cells in bright-field microscope images. In: International Conference on Mass Data Analysis of Images and Signals in Medicine, Biotechnology, and Chemistry, pp 158–172. Springer

Turaga SC, Murray JF, Jain V, Roth F, Helmstaedter M, Briggman K, Denk W, Sebastian Seung H (2010) Convolutional networks can learn to generate affinity graphs for image segmentation. Neural Comput 22(2):511–538

Tuytelaars T, Van Gool L (2000) Wide baseline stereo matching based on local, affinely invariant regions, pp 412–425. ESAT – PSI, Processing Speech and Images

Tuytelaars T, Van Gool L (2004) Matching widely separated views based on affine invariant regions. Int J Comput Vis 59(1):61–85

Vincent L (1993) Morphological grayscale reconstruction in image analysis: applications and efficient algorithms. IEEE Trans Image Process 2(2):176–201

Viola P, Jones M (2001) Rapid object detection using a boosted cascade of simple features. In: Proceedings of the 2001 IEEE Computer Society Conference on Computer Vision and Pattern Recognition. CVPR, Kauai, HI, USA

Vu N, Manjunath BS (2008) Graph cut segmentation of neuronal structures from transmission electron micrographs. In: Image processing, 2008. ICIP 2008. 15th IEEE International Conference on, San Diego, CA, USA, pp 725–728. IEEE

Walter T, Shattuck DW, Baldock R, Bastin ME, Carpenter AE, Duce S, Ellenberg J, Fraser A, Hamilton N, Pieper S et al (2010) Visualization of image data from cells to organisms. Nat Methods 7:S26–S41

Wdowiak M, Markiewicz T, Osowski S, Swiderska Z, Patera J, Kozlowski W (2015) Hourglass shapes in rank grey-level hit-or-miss transform for membrane segmentation in her2/neu images. In: International Symposium on Mathematical Morphology and Its Applications to Signal and Image Processing, Reykjavik, Iceland, pp 3–14. Springer

Williams DJ, Shah M (1992) A fast algorithm for active contours and curvature estimation. Image Understand 55(1):14–26

Wilson CA, Theriot JA (2006) A correlation-based approach to calculate rotation and translation of moving cells. IEEE Trans Image Process 15(7):1939–1951

Würflinger T, Stockhausen J, Meyer-Ebrecht D, Böcking A (2004) Robust automatic coregistration, segmentation, and classification of cell nuclei in multimodal cytopathological microscopic images. Comput Med Imaging Graph 28(1):87–98

Yang S, Kohler D, Teller K, Cremer T, Le Baccon P, Heard E, Eils R, Rohr K (2008) Nonrigid registration of 3-d multichannel microscopy images of cell nuclei. IEEE Trans Image Process 17(4):493–499

Yoo TS, Ackerman MJ, Lorensen WE, Schroeder W, Chalana V, Aylward S, Metaxas D, Whitaker R (2002) Engineering and algorithm design for an image processing api: a technical report on itk-the insight toolkit. Stud Health Technol Inform 85:586–592

Zimmer C, Labruyere E, Meas-Yedid V, Guillen N, Olivo-Marin J-C (2002) Segmentation and tracking of migrating cells in videomicroscopy with parametric active contours: a tool for cell-based drug testing. IEEE Trans Med Imaging 21(10):1212–1221

Zitova B, Flusser J (2003) Image registration methods: a survey. Image Vis Comput 21(11):977–1000

MIX
Papier aus verantwortungsvollen Quellen
Paper from responsible sources
FSC® C105338

If you have any concerns about our products,
you can contact us on
ProductSafety@springernature.com

In case Publisher is established outside the EU,
the EU authorized representative is:
Springer Nature Customer Service Center GmbH
Europaplatz 3, 69115 Heidelberg, Germany

Printed by Libri Plureos GmbH
in Hamburg, Germany